ALSO BY JAMES TREFIL

Meditations at Sunset

Meditations at 10,000 Feet

A Scientist at the Seashore

The Moment of Creation

The Unexpected Vista

Are We Alone?
with Robert T. Rood

From Atoms to Quarks

Living in Space

The Dark Side of the Universe

Space Time Infinity

Reading the Mind of God

The Dictionary of Cultural Literacy
with E. D. Hirsch and Joseph Kett

Science Matters
with Robert M. Hazen

SHARKS
HAVE NO
BONES

1001
THINGS
YOU
SHOULD
KNOW
ABOUT
SCIENCE

Originally published as
*1001 Things Everyone
Should Know About Science.*

James Trefil

A FIRESIDE BOOK *Published by*
Simon & Schuster
New York   London   Toronto   Sydney   Tokyo   Singapore

FIRESIDE
Simon & Schuster Building
Rockefeller Center
1230 Avenue of the Americas
New York, New York 10020

First Fireside Edition 1993
Published by arrangement with
Doubleday, a division of Bantam
Doubleday Dell Publishing Group, Inc.
666 Fifth Avenue, New York, NY 10103

Originally published as *1001 Things Everyone Should Know About Science*.

FIRESIDE and colophon are registered trademarks of Simon & Schuster Inc.

Manufactured in the United States of America

1   3   5   7   9   10   8   6   4   2

Library of Congress Cataloging-in-Publication data is available.

ISBN: 0-671-79627-5

Contents

To my mother, Sylvia Elizabeth Trefil

INTRODUCTION

Science has given us a view of how our universe works that is unrivaled in its beauty and comprehensiveness—from the explosion of a distant star to the working of every cell in our bodies. The acquisition of this knowledge is surely one of the greatest achievements of the human mind.

Because science deals with so many disparate things, there are many ways of presenting it—many ways of slicing the pie. You can, for example, concentrate on the general principles that underlie the physical world. You can look at a specific field—astronomy, molecular biology, geophysics—in great detail and ignore its links to the other parts of the world. Or you can, as I do here, break all of science down into bite-sized chunks. Each of these approaches is appropriate in its own context.

What I try to do in the following pages is give an overview of how the world operates. The information is presented in a numbered list of "things," ranging in length from a single sentence to a few paragraphs. These items, in turn, are grouped together into sections of a few dozen items that cover specific fields of knowledge such as classical genetics, quantum optics, and animal reproduction. Within each section there is a logical progression from the first item to the last, but you can also use the book by turning to the general, classical biology, evolution, molecular biology, classical physical science, modern physical science, earth science, and astronomy—in whatever order you wish.

The book, in short, is intended to be browsed. You are supposed to open it to a random page, read a bit, say "Gee, I didn't know that" or "How interesting," and then put it down until next time. It's not a textbook, and *you aren't supposed to read it from start to finish.* If something catches your attention, read on. If it doesn't, try something else.

Inevitably, putting scientific knowledge together in this unconventional way raises questions, both for the author and the reader. Not every fact about the world is equally important. The first law of thermodynamics (item 535) is surely higher up in the scheme of things than the fact that sharks have no bones (item 25). Everyone really needs to know the first law of thermodynamics (otherwise known as the law of the conservation of energy) to understand the world, but shark anatomy is just one example among many that illustrate the complexity and diversity of biological systems.

In the same vein, you will find occasional "awards" for the achievements—sometimes dubious—of various scientists, "pop quizzes" to keep you awake, and the occasional dumb question that occurred to me while thinking about some field. So don't expect to find that item 20 is ten times more important than item 200, or that the item numbered 1 is the most important of all. In fact, to my mind the most important item in the list is number 1001, for reasons that will be obvious when you get to it.

There is also the question of boundaries: where do you draw the line? The physical world is incredibly rich, and trying to capture it all in a small number of concise statements isn't easy. (Indeed, the first draft of this book had over 1,500 "things," and the process of trimming it down was painful, at least to the author.) Consequently I decided to confine my attention to the traditional natural sciences, and leave both medicine and technology for later books in this series.

Finally, you can ask why I chose to use 1001 things. Well, why not? It's as good a number as any, and there are certainly impeccable literary precedents for it. And while I don't flatter myself by imagining that any of my items has the power or beauty of Scheherazade's stories, taken together my 1001 bits of fact, theory, philosophy, and history will fill out your view of the world and provide, I hope, the bits and pieces of knowledge that you never knew you needed.

James Trefil

Fairfax, Va.

1

CLASSICAL BIOLOGY

A leopard on the Serengeti plain.

Plant Reproduction

1 **Plants can reproduce sexually or asexually.** When the crabgrass in your lawn sends out feelers that take root, it is reproducing asexually. It does this in addition to (and sometimes instead of) the more normal sexual reproduction that takes place through the use of seeds (see below). Bulbs, and rhizomes (underground stems, sometimes called suckers) are other examples of asexual reproduction in plants. The practice of grafting (joining the branch of one plant to the stem of another) is an example of an artificially induced asexual means of reproduction.

The simplest form of asexual reproduction is practiced by single-celled plants like algae, which reproduce by ordinary cell division.

An asexually reproduced plant is genetically identical to the parent plant, and is therefore a clone. Asexual reproduction proceeds more rapidly than sexual, but produces a population in which variation occurs only through mutations.

2 **Alternation of generations is the most primitive form of sexual reproduction.** Plants like ferns and mosses (as well as fungi) use a reproductive technique

ANATOMY OF A FLOWER

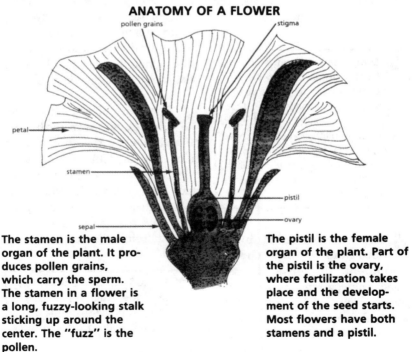

pollen grains

stigma

petal

stamen

sepal

pistil

ovary

The stamen is the male organ of the plant. It produces pollen grains, which carry the sperm. The stamen in a flower is a long, fuzzy-looking stalk sticking up around the center. The "fuzz" is the pollen.

The pistil is the female organ of the plant. Part of the pistil is the ovary, where fertilization takes place and the development of the seed starts. Most flowers have both stamens and a pistil.

in which two separate, alternating life forms are present. Take ferns as an example. The large frond grows from an embryo by ordinary cell division. Spores develop in little pods underneath the fern leaf. These spores each have half the normal complement of chromosomes. When the spores are scattered, they grow into microscopic plants that produce either eggs or sperm. When the sperm mature, they are released and swim through water to fertilize the eggs in neighboring plants. The fertilized embryo, which now has a full complement of chromosomes, then develops into the familiar large fern frond and the whole cycle repeats. For the fern, one generation (the frond) is large and long-lived, while the other is ephemeral and tiny—but both are necessary for the life cycle of the plant.

3 **Every navel orange comes from a single tree.** In the early nineteenth century, a mutant tree appeared on a plantation in Brazil. It produced oranges without seeds. Every navel orange in the world today comes from a bud that was grafted from that mutant onto another tree, whose branches were then grafted onto another, and so forth.

4 **The emergence of plants onto land required the development of seeds.** In seed plants, the egg remains within the parent plant, to be fertilized by sperm that may come from that plant or from another. The fertilized egg (zygote) remains in the parent plant until it develops into a hardy multicelled seed, which is then released to produce a new plant. Nowhere in this process are sperm required to swim through water.

5 **Pollen grains carry the sperm for seed plants.** Inside all that stuff that makes you sneeze and gives you watery eyes every summer are sperm which, if the pollen happens to land near the egg on the appropriate plant, will fertilize the egg and cause the seed to start growing. To propagate itself, then, a plant has to find some way of getting the pollen to the ovary.

By far the simplest way to do this is self-pollination—the pollen moves from the stamen to the pistil without leaving the flower. In cross-pollination, pollen from a different plant fertilizes an egg. The flower can be carried from one plant to another by the wind or by animals such as bees or hummingbirds. Pollination results in the production of fruit.

6 **The fruit of any flowering plant develops from the fertilized ovary.** The fruit can be something juicy, like a pear, but it doesn't have to be edible from a human point of view. The white puffs on a dandelion and the helicopter-like things that fall off maple trees are both fruit in the technical sense.

7 **The red part of the strawberry isn't the fruit.** It's actually a modified part of the stem. The fruit is the little yellow thing sticking to the side.

The seeds of a strawberry.

Plant Growth

8 **The first step in growth from a seed is germination.** When a seed begins to grow, the first thing it does is take in water from its surroundings. Then a root begins to poke out through the seed coating, followed by a shoot that breaks through the ground and leafs out. It is only after a leaf is ready to operate that chlorophyll forms and photosynthesis starts. Up to this time, the young plant has to live on stored energy in the seed.

Seeds can also stay dormant. Dormancy is a device that keeps seeds from germinating until conditions for growth are right. For example, a seed coat may remain too strong to allow any shoots to break through until it has soaked in water for a while, or until it has been exposed to an extended period of cold. This assures that the plant will only start to grow when conditions are appropriate. In the western United States, for example, grass seeds will not sprout unless there are certain levels of rainfall, a property which allows the plant to "skip" the occasional disastrous dry years.

9 **Weeds often produce dormant seeds.** Some weed seeds will remain dormant unless they are exposed to light or unless the outer coating is damaged. Both these strategies make these weeds more likely to sprout in newly disturbed soil. This is why fresh ground quickly becomes covered with weeds.

10 **Material for the tissue of the plant comes from both the air and the ground.** The carbon and oxygen atoms that are incorporated into all living tissue enter the plant leaves as carbon dioxide from the air. A long list of other essential materials, including nitrogen and minerals like phosphorus, potassium, sulfur, calcium, magnesium, and a number of trace elements, are taken up from the soil by the roots. The plant then converts these inorganic materials into living tissue.

11 **Plants cannot use nitrogen directly from the air, where it appears in the form of nitrogen molecules** (N_2). Plants can only use nitrogen if it has been "fixed," or converted into ammonia (NH_3). The fixing of nitrogen is carried out by single-celled organisms, some of which are monera and some of which are algae.

Without these so-called nitrogen fixers, no higher life could exist on earth. Without them there would be no multicelled plants, hence no animals and no human beings.

In the ocean, nitrogen is fixed by algae and photosynthetic bacteria. In the soil, there are a few free-floating nitrogen-fixing bacteria, but most of the nitrogen is fixed by bacteria living in nodules of the roots of plants. Peas, soybeans, and alfalfa are a few plants that harbor nitrogen-fixing organisms in their roots.

12 Crop rotation introduces fixed nitrogen into the soil. Farmers have known for centuries that planting crops like alfalfa or clover on a field enriches the soil and makes subsequent plantings of other crops more productive. This system works because these particular crops harbor nitrogen-fixing bacteria in their roots, and the bacteria fix much more nitrogen than is used by the plant. The excess (together with the material from the plant roots themselves) forms a "capital" of fixed nitrogen in the soil that other plants can use.

13 Some plants have evolved themselves into strange little ecological corners. Mistletoe, for example, is a parasite. It supplies some of its needs through photosynthesis (it is green, after all), but drains nutrients from the trees on which it grows.

Similarly, plants like the Venus's-flytrap enjoy an occasional snack in the form of an insect to supplement their photosynthetic output.

Animals

14 The kingdom of animals ranges from primitive structures like sponges all the way up to human beings. It is the most diverse of the five kingdoms. Sponges contain many cells, but each cell can operate independently of the others—if you pass a sponge through a sieve, for example, each cell is capable of producing a new organism. In higher animals, like humans, cells have specialized functions and depend on each other for survival.

15 Animals ingest their food. This, in fact, is the great evolutionary strategy of the animal kingdom. Unlike plants, which manufacture their own food through the process of photosynthesis, animals have to take their food from the environment. They can do this using one of two strategies: they can sit still and let the food come to them (like corals) or they can go out and look for it (like leopards).

Herbivores (like rabbits) get their

food by eating plants, carnivores (like wolves) get it by eating other animals, and omnivores (like humans and raccoons) eat both plants and animals.

16 **There are many phyla of animals.** The animal kingdom is divided by some biologists into as many as thirty-one different phyla. Most of these include lower forms such as worms and parasites. You can get some of the flavor of how these classifications work by looking at this representative (but not exhaustive) list:

Porifera (sponges)
Cnidaria (jellyfish, corals, and sea anemones)
Plathelminthes (flatworms, including tapeworms)
Rotifera (microscopic organisms)
Nematoda (roundworms)
Annelida (segmented worms)
Mollusca (clams)
Arthropoda (spiders, insects, crustaceans)
Echinodermata (starfish)
Chordata (everything with a spinal chord, including humans)

17 **Animals evolved through a number of primitive forms.** The ancestors of higher animal forms include things like roundworms, flatworms, segmented worms, jellyfish and corals, and mollusks. Each of these represents a separate phylum of animal life that biologists consider just as worthy of study as our own phylum of chordates.

Looks can sometimes be deceiving. Animals like starfish and sea urchins, though they look simple, are pretty complex organisms. In fact, they represent the last branching of the evolutionary tree which leads to chordates, vertebrates, and, ultimately, to us.

18 **The most successful phylum of animals consists of the arthropods.** This phylum includes spiders, centipedes and millipedes, crustaceans such as lobsters, and, most important of all, insects. Arthropods are characterized by a hard exterior coating (exoskeleton), which is usually jointed to allow movement. The exoskeleton

Sponges (top) and coral (bottom) are two different kinds of animal.

does not grow, so arthropods must shed their old exoskeleton (molt) periodically as they grow. From 50 to 80 percent of all species on earth today are arthropods.

The horseshoe crabs whose shells wash up in abundance on East Coast beaches are arthropods who have survived virtually unchanged for almost 500 million years.

19 The strangest animal has its own phylum. Deep under the surface, clustered around hydrothermal vents in the ocean floor, lives one of the strangest animals known. A reddish worm that creates a long, tough tube to live in, it ranges up to 25 feet long, ingests its food, but has no organs that correspond to a mouth or intestines. Apparently these worms are nourished by bacteria that live inside their cells. They have an entire phylum all to themselves, since no other animal is at all like them.

Giant tube worms living in the Galápagos Rift at the bottom of the Pacific Ocean.

20 Centipedes don't have 100 legs. Different species of this class of arthropods have between 15 and 173 pairs of legs. Species of millipedes (another class of arthropods) have between 20 and 400 pairs.

How many legs do you count?

21 Insects are the most successful arthropods. Estimates of the total number of insects on the planet run up to 10^{18}— roughly a billion for every human being. They all have three pairs of legs (which differentiates them from spiders, which have four), an exoskeleton, and three parts to their bodies—a head, a thorax, and an abdomen.

22 "God has an inordinate love of beetles." With this comment, the eminent British biologist J.B.S. Haldane is said to have answered a questioner who wanted to know what studying nature had taught him about the mind of the Creator. Among the successful insects, the most successful order is Coleoptera, or beetles. There are more species of beetles than of anything else on earth. I was an avid insect collector as a boy, and I was astonished to learn that there were hundreds of different species of beetles in the Chicago area. I found the task of collecting them all rather daunting, and switched to physics before I finished.

Enduring Mystery

23 **Where did the vertebrates come from?** Tracing out the evolutionary steps that led to present-day vertebrates turns out to be rather difficult. One current theory about how animals acquired a backbone, based on the observation of living animals, is this: there are some animals whose larvae swim around and have something like a spinal chord. In this stage, they resemble primitive tadpoles. In adult life, however, they lose both the ability to move and the spinal chord. The theory: animals like these evolved to a stage where the adult phase was eliminated—in effect, they lived all their life as larvae. Once this step was taken, the bony protection for the (now vulnerable) spinal chord evolved, and vertebrates were off and running.

24 **There are many orders of vertebrates.** Vertebrates are the most familiar animals, of course, and there are many orders within the subphylum. The orders are:

fishes (three different kinds)
amphibians
reptiles
Aves (birds)
mammals

25 **The shark has no bones.** Its entire "skeleton" is made from cartilage, a fact that explains its flexibility when swimming. Sharks and lampreys are surviving representatives of the most primitive forms of fish.

As early as 400 million years ago, the oceans teemed with fish, which were the most advanced form of life at the time. Many of these early fish already had bony skeletons. Many were huge things with armored heads and bodies, and are now extinct. Such bony fish (which now include everything except sharks and lamprey) evolved in fresh water and only later moved to the ocean.

26 **Some primitive fishes had lungs that breathed air.** The first bony fishes had lungs, probably to help them get more oxygen. In most fish, these lungs evolved into swim bladders and are no longer used for respiration. It used to be thought that oceangoing lungfish had been extinct for over 70 million years, but in 1939 fishermen pulled a living specimen from the Indian Ocean. Since then, many more have been found, and there's no doubt that at least one "fossil" animal is still around today.

Oddly enough, what nature couldn't do, human beings may accomplish. The "living fossils" of the Indian Ocean are so valuable as museum specimens that they may well be hunted to extinction by native fishermen.

27 **Amphibians such as frogs and salamanders**

evolved from lungfish. They carry a reminder of their origin in the fact that they still spend part of their life cycle in water. The crucial step in the move to land was the evolution of fins into legs capable of moving the animals around in the new environment.

28 **Reptiles were the first vertebrates completely adapted for life on land.** They include turtles, snakes, lizards, and crocodiles. The features that distinguish them from amphibians are scales (to control water loss), a tough egg with a lot of yolk (to allow the young to grow before they emerge), a heart capable of moving oxygen around the body more efficiently, and a more complex brain.

29 **Birds descended from reptiles.** Birds are characterized by having feathers (which evolved from scales), a heart with a two-chambered ventricle, and a larger brain than reptiles. They also have a large bone in the chest where the muscles used in flying (the "white meat") is attached. All birds were originally equipped for flight, although some (such as the ostrich) have evolved away from that mode of life.

30 **Birds are warm-blooded.** Amphibians and reptiles are cold-blooded—that is, their body temperature depends on the temperature of the surrounding air. This is why snakes and frogs are so sluggish in the morning and why they spend so much time sunning themselves.

Birds, on the other hand, have a metabolism that maintains a constant body temperature. They are the lowest form of life for which this is clearly true, although there is debate over whether some dinosaurs were also warm-blooded.

31 **Mammals are animals that have hair, large brains, and suckle their young.** They are also warm-blooded. Human beings belong to the class of mammals, as do most familiar large animals. Warm-bloodedness allows mammals to function in cold climates, where cold-blooded animals cannot survive, while the large brain allows them to adopt various social strategies not available to other life forms.

Mammals were around throughout the "Age of Reptiles"—they did not appear suddenly when the dinosaurs were wiped out. They played a minor role in ecosystems throughout the Mesozoic era in the form of mouselike creatures eking out an existence in a world populated by giant reptiles. They blossomed only after their competition was removed.

How Animals Are Put Together

32 **Every animal is the sum of its organ systems.** The cells that exist in your body (and the body of any animal) are not scattered at random. They are arranged in organs (like the stomach and heart), and the organs are themselves arranged into organ systems (like the digestive and circulatory systems). The composite of these systems makes up the complete animal.

Digestive Systems

33 **The digestive system converts food to materials the cells can use.** Animals take in food, either plant material or meat. The food is passed into the digestive system (usually a hollow tube that runs through the body) where the large molecules are broken down into smaller molecules by the actions of enzymes and then absorbed into the animal's body. The organs that accomplish this task make up the digestive system.

34 **In humans, digestion starts in the mouth.** When you chew your food, you are starting the digestive process by breaking it up into small pieces (your mother was right about chewing!). Meanwhile, an enzyme in the saliva starts to break down starch.

35 **Human digestion continues in the stomach** and intestines. In the stomach, hydrochloric acid kills microorganisms and helps to produce an enzyme (pepsin) that starts to break down proteins. In the small intestine, enzymes made in the intestine wall, the liver, and the pancreas carry out the major task of breaking down carbohydrates, proteins, fats, and nucleic acids. The products of this breakdown are absorbed through the walls of the small intestine. In the large intestine, water is taken from the digested foods before it is excreted.

36 **Many symbiotic bacteria live in the human large intestine.** The most famous of these is *Escherichia coli* (*E. coli* for short). A large part of our knowledge of molecular biology comes from experiments done on strains of this bacterium grown in the laboratory.

37 **Cows "ruminate" while chewing their cud.** Like many animals, the cow has nothing in its digestive tract to allow it to digest cellulose directly. Instead, chewed-up grass goes to a chamber in front of the true stomach, called the rumen, where resident microorganisms start to break the food down. While this process is going on, the cow periodically brings the material back up to be chewed. In the end, the digested grass (now in the form of fatty acids) and the microorganisms pass through a true stomach and are digested.

38 Cows aren't really vegetarians. Grass isn't a good source of protein, so the cow gets double duty from its microorganisms. First, they break down the grass, then, this job completed, they make the ultimate sacrifice and are themselves digested by the downstream part of the cow's stomach.

Sensory Systems

39 Animals learn about their environments by way of their sensory systems. In general, these systems respond to one of four things: light, mechanical pressure, temperature, and chemical concentrations. The five senses of the human being, for example, are sight (detecting light), smell and taste (detecting chemicals), and touch and hearing (sensing pressure and vibrations). We have no specialized temperature sensor. I have often wondered if this explains the fact that thermometers weren't developed until well into the seventeenth century.

40 Animal eyes can be simple or complex. Some single-celled protists have patches on their outer surfaces that allow them to tell light from dark. This allows them to swim toward the light (i.e., the surface) on the ponds in which they live.

Insects and humans, on the other hand, have very complex eyes. The eye of insects (and other arthropods) is made up of multiple lenses stacked together. Each piece of the compound eye is, in reality, a separate "mini-eye," complete with a lens to focus light on an individual receptor. The insect sees the world, then, as a series of overlapping patches. It can't see the kind of fine detail we can, but it can detect motion more efficiently.

41 Dragonflies may have over 20,000 lenses in *each* eye. (I'd like to know who did the counting!)

42 In humans, and most vertebrates, the eye is quite complex. Light enters through the opening of the pupil (the dark part of your eye) and is focused by the lens. Muscles in your eye squeeze or relax to change the focal length of the lens, thus allowing you to see things at different distances from you. The light is focused

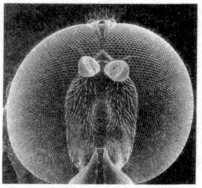

An electron microscope photograph of the eye of a syrphid fly, magnified 120 times.

on the *retina* at the back of the eye where chemical reactions (see below) produce a nerve impulse that travels over the *optic nerve* to the brain.

There are two kinds of light-sensitive cells in the eye—the "rods" and "cones"—with names having to do with the shape of the cells, not their function. The rods respond to very low levels of light, but not to color. They allow you to see in the dark. There are three kinds of cones, sensitive, respectively, to blue, green, and red light; the cones give us color vision. In both rods and cones there are large molecules that absorb photons of light and eventually produce impulses in the optic nerve.

43
Frogs, birds, lizards, and humans can see colors, but dogs cannot. The way you see colors is, in fact, a lot more complex than you might think. Color depends on the ambient light—this explains why clothes often look different under the fluorescent lamps in a store than they do in sunlight. But it also depends on the way color is processed in the eye (painters have known for a long time, for instance, that if you put blue next to yellow, the area around the boundary looks whitish and washed out). Finally, it depends on your mindset. People shown a tree and a brick in the same shade of gray will see the brick as more reddish and the tree as more greenish.

44
The eye is not like a TV camera. A TV camera produces an image by converting what it sees into a series of bright and dark dots (or, in the case of color TV, three sets of dots—one for each primary color). There is a one-to-one connection between each spot on the object being seen and each spot in the final image, and this connection is maintained all they way through the process by which the camera produces it. You could, in other words, interrupt the process at any point and say something like "This electronic signal is from that particular spot of that particular leaf."

In the brain you can't do this. A piece of the visual cortex may connect to many different parts of the retina, and the process of "seeing" is quite complex. There appear to be, for example, parts of the cortex that are very good at recognizing horizontal lines, others that recognize vertical lines, others that recognize edges of objects, and so on. This complex architecture is what makes the brain so much better at processing visual information than even the fastest computer, which, like a TV camera, has to work with information on a one-to-one basis.

45
Ears respond to the pressure exerted by sound waves. In the human ear sound waves make the eardrum vibrate back and forth like a drumhead. This motion is carried through a series of small bones to the inner ear, where they cause pressure changes in a liquid in the spiral-shaped organ called the *cochlea*. The pressure has the effect of causing certain

sensitive cells to deform, and this in turn produces a signal that eventually is transmitted to the brain.

46 **Not all animals have their "ears" on their heads.** Some moths have the equivalent of the eardrum mounted in the middle of their thorax, while spiders and crickets have them on their legs.

47 **Taste and smell involve chemical receivers.** In order to taste something, molecules from the material being tasted have to come into contact with specialized cells that are part of the taste buds on the tongue. In order to smell something, molecules of the material being sensed must travel through the air to your nose, where they interact with other specialized cells. In both cases, the reaction of the molecules with the cells produces signals that travel through the nervous system to the brain.

The dog's legendary sense of smell is reflected in the animal's anatomy. A dog has over 200 million olfactory cells in its nose, while a human has only 5 million. What are we missing?

48 **The female of the silkworm moth announces her availability for mating by giving off a substance called bombykol,** which is the ultimate perfume. Male moths can "smell" bombykol when it is diluted to one molecule in a quadrillion in the air—probably the most prodigious feat of chemical sensing in the animal kingdom.

49 **The sense of touch involves many different receptor cells.** There are specialized cells near the surface of the skin that signal pain and others that respond to the pressure of a touch. Deeper down there is a network of different cells that perform the same function. Finally, there are cells wrapped around hairs that tell you when the hair is moved.

Houseflies have pressure-sensitive cells on their bodies that tell them when a large body is descending on them, pushing the air aside as it goes. This is why it's so hard to swat a fly with your hand and why fly swatters have holes in them to let the air through.

Bones and Muscles

50 **Every multicelled animal has to have a way to support itself against the pull of gravity.** The most common solutions to this problem are to put a skeleton on the outside (clams and insects) or on the inside (humans). These two strategies involve, respectively, structures called an exoskeleton and an endoskeleton.

51 **Vertebrate endoskeletons contain bones and cartilage.** The bone appears where stiffness and load-bearing capability are required, and cartilage ("gristle") in places that require resilience. Your nose and voice box are made of cartilage, for example, and the material serves as a shock absorber in the joints.

In vertebrate skeletons, joints are held together by ligaments. These are tough, relatively inelastic bands that connect bones on one side of a joint to bones on the other—like cables. The relative inelasticity of ligaments and their slowness to heal explains why knee injuries are so calamitous to the careers of athletes.

52 **There are two kinds of muscle in the human body.** Muscles are made of bundles of long cells that contract upon receiving the proper signal from a nerve. In the human body the simplest of these are *smooth muscles,* which control involuntary motions like the dilation of the pupil and the contractions of the stomach and intestine. *Striated muscles* are the ones

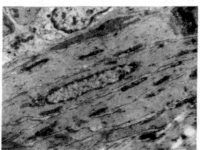

Examples of striated muscle (top) and smooth muscle (bottom) as seen under a microscope.

that make us move. They have a more complex structure than (and evolved later than) smooth muscles. A special subset of striated muscles beat rhythmically to pump blood through the heart.

Pop Quiz

When you tell a kid to ''make a muscle,'' what kind of muscle do you mean? Answer: Striated.

53 **Muscles are attached to bones by tendons.** When a muscle contracts, it pulls on the tendons, which pull on the bone and cause movement. "Tennis elbow" is a common affliction that involves inflammation of the tendons in the elbow.

Any repeated strain on the elbow will produce this condition. For example, I contracted "tennis elbow" by using a chainsaw to cut firewood.

Nervous System

54 **An animal's nervous system collects, processes, and acts on information.** The *sensory nervous system* takes information from the animal's sense organs and feeds it into the *central nervous system* (in humans, the brain and spinal cord), where it is processed. Once an appropriate response to the environmental situation has been determined, other signals from the central nervous system go out through the *autonomic nervous system* (which controls involuntary actions like the beating of the heart) and the *somatic nervous system* (which controls voluntary actions like moving a limb).

55 **Signals in the nerves are different from ordinary electrical current.** Signals travel through the nervous system along nerves, which are networks of individual cells called neurons. Each nerve can carry many signals at the same time, much as a telephone cable can carry many independent conversations. The points of communication between the ends of individual neurons are called synapses and the long thin parts of a nerve cell that actually carry a signal (the "wires") are called axons.

In electronic systems like your stereo set, signals are carried through the wires by the movement of electrons. In nerves, the signals involve the movement of ions of potassium and sodium across the outer membrane of the axon. When a signal arrives at one end of a neuron, chemicals called neurotransmitters are sprayed out and picked up by the next neuron and the signal moves on. A typical response time for a neuron is a millisecond (.001 seconds), which is over a thousand times slower than the response time of the equivalent component in your personal computer.

56 **The nervous system of higher animals is concentrated.** Among animals like jellyfish, the nervous system is spread around the body, more or less like a system of roots. In higher forms,

both the sense organs and the parts of the nervous system that process and act on sense information are in the head. Sometimes, as in worms, this central processing system is just a bundle of nerves (called a ganglion). In vertebrates, however, it becomes the complex structure we call the brain.

57 **Different parts of the human brain carry out different functions.** You can divide the brain (roughly) into three sections. The hindbrain is located at the base, where the spinal cord comes in. It governs things like automatic motor movements—when you move unconsciously to regain your balance, for example, the neural signals came from this part of the brain. The outer covering of the brain ("the gray matter") is the cerebral cortex, and this is where sense data are processed and where higher functions like reasoning and memory are located. In between the two is the midbrain, where emotions and some behavior come from.

58 **Scientists do not fully understand the brain's complexity.** There is a tendency in popular books to make facile identifications of the three parts of the brain: hindbrain = reptilian, unconscious existence, midbrain = animal emotions, cortex = "higher" functions. Or you might have run across the identification hindbrain = id, midbrain = ego, cortex = superego.

It's not that simple! Sorting out the complexities of the human brain is likely to keep scientists occupied for a long time to come, and over-simplified ideas like this are no longer accepted by researchers.

59 **In addition to signals traveling along the nerves, animals control their body functions through the use of hormones.** These molecules are secreted by specialized glands or tissues and then travel around the body and affect other organs. In humans, these glands constitute the *endocrine system*. For example, if you are frightened, glands on your kidneys secrete adrenaline, which causes the heart to speed up, and increases the flow of blood to the muscles.

Circulation, Respiration, and Excretion

60 **The basic chemical reaction that supplies animals with energy is oxidation (burning).** In order to make this reaction go, there has to be a way to get oxygen into the animal's body and to the cells, and then get the waste products away from the cells and removed from the body. Three separate operations combine to perform these tasks: respiration (getting the oxygen), circulation (getting oxygen to the cells and wastes away from them) and excretion (removing wastes from the body).

61 **The way an animal gets oxygen from the environment depends on its size**

and on whether the animal lives in water or air. A one-celled organism can get enough oxygen in (and CO_2 out) by having the gases diffuse though its outer surface. It needs no respiratory system. In gills (both in arthropods and fish) water is pumped continuously over surfaces containing blood vessels, and oxygen diffuses across to the blood (and CO_2 diffuses out) "on the fly." Lungs are an adaptation for life on land. In lungs, air is drawn into a sac and allowed to sit while the exchange of gases goes on. Insects do not have lungs but a series of tubes (called trachea) that carry air from holes in their bodies directly to their cells.

62 **Warm-blooded aquatic animals that need a lot of oxygen because of their size cannot get enough from water.** A given volume of water contains only a few percent of the oxygen of the same volume of air, so whales and porpoises breathe air. Also, water loses its ability to hold oxygen as it warms up, so warm water has even less oxygen than cold. This is why fish move to deep (and cold) holes during the day.

63 **Advanced animals have hearts.** The function of an animal's circulatory system is to move oxygen and nutrients to the cells and to remove waste products from them. In simple animals (like nematodes) the blood just sloshes around in an interior cavity. In more advanced animals it is pumped around the body by a heart.

A heart consists of two kinds of

chambers—one where blood is gathered (called an atrium) and one where it is pumped out (called a ventricle). In fish, where blood makes one pass through the gills and then goes directly to the cells, the heart is composed of two chambers—one of each kind. In humans, there is one set of chambers to pump blood to the lungs, another to pump it through the body, so the human heart has four chambers all told.

64 Blood flows from the heart in arteries, to the heart in veins. In humans, blood is pumped from the left ventricle into a system of branching arteries from which it eventually reaches tiny vessels called capillaries located throughout the body. In the capillaries the oxygen moves into the cells and CO_2 and other wastes enter the blood, which then flows back to the heart in the veins. After being collected in the right atrium, the blood is pumped from the right ventricle to the lungs, where it dumps the CO_2 and picks up oxygen. From the lungs it goes back to the left atrium and starts the whole process over again.

65 William Harvey (1578–1657) discovered the circulation of the blood. Before Harvey's work was published in 1628, the role of the heart in circulation was not recognized—indeed, for most of human history people thought the blood didn't move at all. In a series of classic experiments, Harvey established our modern picture of the circulatory system. A typical experiment: he put a tourniquet on someone's arm, then, after the veins had popped, pressed on them to see which direction was "downstream." This is how he discovered that blood in the veins always flows toward the heart.

66 The pressure exerted by the heart is not enough to push blood back through the veins. This is particularly true if it has to be pumped uphill. The normal movement of body muscles acts to push the blood through the veins, which also contain valves to prevent backflow.

67 Blood is an extremely complex substance. Over half of it is a yellow fluid called plasma, which carries most of the chemical nutrients. The red blood cells transport oxygen and the larger but less numerous white blood cells create a defense against foreign bodies and microorganisms. The final component of the blood is platelets, fragments of bone marrow cells that help blood to clot.

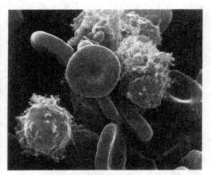

The round, doughnut-shaped objects are human red blood cells; the puffy white objects are white blood cells.

A moderately complex molecule called hemoglobin makes the blood red. A diagram of its atomic structure looks something like a lace doily. At its heart is a single atom of iron that can loosely bond to four atoms of oxygen. Hemoglobin allows the blood to carry about four to seven times as much oxygen as it could carry were the oxygen simply dissolved in the plasma.

68 Red blood cells do not divide. They are created in your bone marrow at the rate of 140,000 per minute and, after a useful life of a few months, destroyed in the liver.

69 Every animal must have a way to remove waste materials. There must be some way to sort through body fluids to get rid of unneeded or harmful materials, and there are many different ways that animals perform this function. In simple animals (such as flatworms), there are single cells open to the outside. As fluid moves through the cells, some materials are reabsorbed—the remainder is excreted. In others (crustaceans, for example) blood is processed by filtration. You may have seen the green organ where this occurs the last time you ate lobster—it's considered quite a delicacy.

70 In vertebrates, waste disposal takes place in the kidney. Blood enters the kidney and is processed in a collection of small organs called nephrons. The technique is simple: the blood is filtered, then materials are selectively reabsorbed. What isn't reabsorbed passes out of the kidneys to the urinary bladder, from which it is excreted. The kidneys are actually a complex chemical factory which maintains a balance of various molecules and water in the body.

The kidneys are very good at maintaining your body's water balance—you can drink as little as a quart a day or as much as several gallons at one sitting and still survive. But the kidneys can't produce urine with a salt concentration of more than 2 percent. If you drink something like seawater (which is 3 percent salt), the kidneys have to remove water from your body to dilute the extra salt, a process that makes you even more thirsty. This explains the origin of the line "Water, water, everywhere/Nor any drop to drink" from the "Rime of the Ancient Mariner."

71 The white part of bird droppings is the "urine." In humans, of course, liquid waste is collected in the bladder and excreted as liquid urine. In insects, reptiles, and birds, on the other hand, water is removed from the urine, the resulting uric acid is mixed with solid wastes, and the two are excreted together. These animals do not urinate.

Animal Reproduction and Development

72 **Animals can reproduce either sexually or asexually.** In sexual reproduction, an offspring has two parents, each of which contributes half the genes. In asexual reproduction, a single parent contributes all the genes. Single-celled organisms reproduce asexually.

73 **Many higher animals reproduce asexually.** Although most higher animals have two parents, some very advanced animals are capable of reproducing asexually. Sponges, for example, commonly reproduce by growing buds on a parent organism. The bud then splits off and grows independently. Some advanced animals, like starfish, can regrow from a single part that has been cut off. Snakes, however, cannot do so, even though there is a common folk belief that they can.

74 **When animals reproduce sexually, two individuals each contribute half of the genes to the offspring.** In sexually reproducing animals, there is always some mechanism by which cells with a full complement of genes can divide to produce cells with only half the number of normal genes. These sorts of cells are called gametes. The male gametes are sperm, the female eggs. Each parent contributes one of these specialized cells to an offspring, so that the off-

This starfish is regenerating two limbs it has lost.

spring has a full complement of genes (half from each parent).

75 **Sexual reproduction need not involve sex.** It is not necessary that two animals mate in order for sexual reproduction to take place. In fact, there are many strategies for bringing the two sets of gametes together. These may include sexual activity as normally understood among humans, but may also involve something as impersonal as having a male and a female each produce large numbers of cells and scatter them to the winds (or, more usually, the currents) in the hope that they will encounter the appropriate matching cell.

Reproductive Systems

76 **The first step in animal reproduction is the development of gametes.** In every sexually reproducing animal, there are specialized cells (called germ cells) that produce either an egg (female) or a sperm (male). The eggs are customarily produced in organs called "ovaries" and the sperm in organs called "testes." These organs may or may not occur together in the same animal. Sponges, flatworms, and mollusks normally contain both male and female organs. In humans, of course, sexual organs appear in different individuals.

77 **Germ cells in the male organs divide by mitosis to increase their number, then by meiosis to produce the sperm.** The final product of this process is a complex structure called a spermatid, in which the head contains the DNA and the long tail provides propulsion.

78 **Antonie van Leeuwenhoek (1632–1723) was the first person to see human sperm** and to understand their role in reproduction. He believed, however, that the head of each sperm contained a miniature human being, which would grow to maturity after fertilization.

79 **Ex Ovo Omnia.** "From the egg, everything." This is how William Harvey (1578–1670) summarized the discovery of the role of the egg in reproduction. This discovery ended a long scientific enquiry into the precise mechanics of human reproduction.

Once the germ cells that produce the egg undergo meiosis, the egg undergoes further development in some animals. This can include the development of a yolk (which carries food for the newborn individual) and even a shell.

The final ovum can vary in size from one species to the next. In humans, for example, the egg is only a little more than a tenth of a millimeter in diameter. Despite this small size, the human egg contains almost two hundred thousand times the volume of a sperm.

80 **Champion ovum.** The largest ovum produced in any animal is about seven inches long, and is seen in some species of sharks.

---

81 Fertilization is the central act of sexual reproduction. It is the process of bringing an egg and a sperm together.

Animals that do not move generally broadcast the egg and sperm into the environment, hoping that some fortuitous juxtapositions will occur. This strategy is called "spawning" and is practiced by animals like oysters and even some fish such as salmon.

Some animals, like frogs, mate by having the male and female eject sperms and eggs into the environment simultaneously. Both this and spawning are examples of external fertilization—that is, fertilization in which the sperm and egg come together outside the body of the female.

A human egg surrounded by spermatozoa. Eventually, one will enter the egg and fertilization will occur.

82 Most advanced species reproduce by internal fertilization. In humans and other mammals, as well as most other advanced animals, fertilization occurs when the sperm is injected into the body of the female and travels to the egg.

The human sperm produces a substance that helps to break through the wall of the ovum, but a single sperm cannot produce enough of this material to get through. Consequently, many spermatids must be involved in breaking down the outer wall of the egg before one of their number can actually get through and initiate fertilization.

83 The queen bee mates only once. The queen bee, soon after she becomes mature, leaves the hive and mates once with a single drone, usually at the altitude of a couple of hundred feet and "on the fly." She then stores all the sperm in special organs in her body and uses them to fertilize eggs for a period of many months or even years. Her stored sperm represents the entire genetic capital of a beehive.

84 The fertilized egg, whose genes come half from one parent and half from the other, is called a zygote. Once it is formed, it may be protected, or it may not.

In oysters and other spawning animals, a fertilized egg is more or less left to its own devices. The parent need spend no effort in protecting it, and the reproductive strategy is to fertilize so many eggs that some are bound to survive. In higher animals, however, there are a number of other strategies for protecting the developing organism. It may be put into a hard-shelled egg for protection, it may be grown completely inside the mother's body, as in human beings, or it may be born early and then carried in a pouch, as in kangaroos and opposums.

85 **Starting from a single cell, the zygote develops into an organism that eventually may contain trillions of different kinds of cells.** This growth and diversification is one of the most inspiring (as well as one of the most mysterious) processes in nature. The instruction for development must be contained in the DNA that was originally in the egg and the sperm. Understanding just what these instructions are and how they work remains one of the great tasks of modern science.

When the zygote has developed into a multicelled organism, it is called an embryo.

86 **Ontogeny recapitulates phylogeny—sort of.** Nineteenth-century biologists noted that, as an embryo of an advanced organism grows, it passes through stages that look very much like the adult phase of less advanced organisms. For example, at one point the human embryo has gills and resembles a tadpole. In the nineteenth century, this so-called biogenetic law was taken to prove that evolution had proceeded on more or less a straight line from the simplest organisms to its epitome in human beings. We no longer have this view

of evolution, but the biogenetic law remains a useful generalization about the way an embryo develops.

87 **The embryo begins to develop through a process called cleavage.** Starting with a single cell, the embryo divides first into two cells, then into four, then into eight, then into sixteen, and so on. For the first few divisions, the cells remain synchronized—meaning they all divide at roughly the same time. Later on, as the cells begin to diversify, this synchronization disappears. If one traces the history of a cell in the early embryo, one finds that some cells will become part of the nervous system, others part of the digestive system, others part of the skeletal system, and so forth.

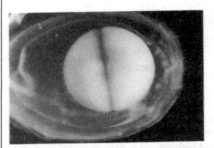

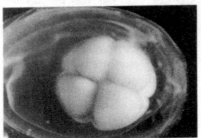

The first two splittings of the salamander zygote—first into two cells (top), then into four (bottom).

88 **The capacity of a cell to change its fate vanishes very early in embryonic development.** The ability of a cell to change its fate (i.e., to change from being part of the skeleton to part of the nervous system and so on) is called its "potency." The potency of a cell vanishes early in the development of the embryo. In later stages of development, if a cell is transplanted from one place in the embryo to the other, it will still develop into the type of organ appropriate to its original position.

The way a cell develops seems to depend primarily on where it is in the embryo—that is, if we take a cell from a region that would normally develop into muscle and move it to a region that will develop into part of the nervous system early in the development of the embryo, that cell will eventually become part of the nervous system.

89 **Cellular development does not stop with birth.** It is usual to think of cellular development of an organism in terms of the embryo, but cellular development goes on long after birth. Any one who has watched a child grow knows that this is true. In fact, some scientists think that our entire life history, from fertilization to eventual senescence and death is programmed into our genes.

Enduring Mystery

---

90 **Why do we grow old?** Recently, biologists have begun to address this question in a scientific way. There are two major schools of thought about the answer to the question. One is the "accumulated accident" school, which holds that aging occurs because our bodies suffer abuse and punishment during our lifetime. The other is the "programmed senescence" school, which holds that aging is built into our genes. The programmed senescence view is supported by the results of some experiments. It turns out, for example, that cells from a human embryo grown in culture can divide only about fifty times before they die, regardless of how much nutrient is available.

From the point of view of evolutionary biology, programmed senescence makes sense because once an organism is too old to reproduce itself, natural selection will not act to weed out unhealthy members of the species. There is, in other words, no evolutionary pressure to produce a long-lived species. On the contrary, if the energy that goes into graceful aging takes away from reproductive capacity, a long postreproductive life can actually be a negative factor in evolution.

91 **There is no "scientific" definition of where "life" begins.** One of the great difficulties one encounters in the debate on abortion in the United States is the question of where "life" begins. Those who oppose abortion hold that life begins with fertilization, or conception. Those who are in favor of a person's right to choose abortion hold that it begins at a later time. It should be clear, however, that there is no single time in the transition from germ cell to zygote to newborn infant when you can say "life begins here." The process is continuous, and the question of where life begins is something that has to be answered from outside the realm of science itself. In this, it mirrors the old theological debate on the question of when the human being acquires a soul.

Ideas About the Origins of Life

92 All life comes from pre-existing life. This is arguably one of the most important precepts of biology. It tells us that no new living things come directly from inorganic matter. Instead, living things come from other living things. This is obvious in the process of cell division, but it was not nearly so obvious before people knew about the existence of cells, back in the days when biologists thought of the organism (rather than the cell) as the basic unit of life.

93 Throughout most of recorded history, people believed in spontaneous generation. This was the belief that life could arise spontaneously from nonliving materials. If you left a piece of meat out to rot, for example, it would soon be covered with maggots. What more evidence do you need? In the same way, frogs and salamanders were thought to come from slime, fleas from sand, and so on.

The idea of spontaneous generation was a long time dying, and it took a series of experiments over a period of several centuries to lay it to rest. The first was by an Italian physician named Francesco Redi. In 1668 he showed that if meat was covered to keep flies away, maggots did not develop. Thus, flies created the maggots which, in time, developed into new flies—life from previous life. The Dutch scientist Antonie van Leuuwenhoek (1632–1723), using a recently developed microscope, followed the life cycle of the flea and showed that fleas, too, come from other fleas. By the end of the eighteenth century, spontaneous generation of entire organisms had been pretty well discredited.

94 Spontaneous generation of cells was harder to disprove. Proving that spontaneous generation didn't happen at the cellular level was difficult. It wasn't, for example, until 1875 that microscopes had developed to the point that mitosis could actually be seen and described. Thus, the fact that grape juice would ferment into wine whether it was covered with a cloth or not was put forward as evidence of spontaneous generation of the yeast. These notions were finally laid to rest by Louis Pasteur who, in a series of ingenious experiments in the late 1850s and 1860s, showed that the air itself is full of microorganisms capable of producing such effect.

Single-celled Organisms

95 There are two entire kingdoms devoted to one-celled organisms. It used to be that one-celled organisms were classed as either plants or animals, depending on whether they got their energy from photosynthesis or by ingesting food. More recently, it has been realized that these organisms really don't fit very well into the traditional categories and two new kingdoms have been devised for them. These kingdoms are Monera, which is made up of one-celled animals that do not possess a nucleus, and Protista, whose members' cells do have a nucleus. It is generally believed that monera are the more primitive of the one-celled organisms and were the first form of life to evolve.

96 Cells of organisms in the kingdom Monera are rudimentary. Not only are these cells prokaryotic (the DNA lies around loose), but they lack much of the complex structure characteristic of more advanced cells. We think that many of the organelles in advanced cells originated as symbiotic organisms—that is, that advanced cells are actually colonies made of many different simple cells that have learned to live together. Thus, one way of thinking about the kingdom Monera is to suppose that they are what cells were like before all the symbiosis started.

97 Monera are the generalists of the cellular world. Perhaps because they are relatively simple, monera seem to have abilities that more advanced organisms have lost. For example, they can digest cellulose, which mammals can no longer do. This is why cows carry monera along in their stomachs.

I think of monera as being analogous to a simple personal computer that is ready to go every time you turn it on, and more advanced cells as being analogous to sophisticated machines that can do more things, but have to be pampered and reset every time they are used.

Like all other living things, monera need to have a source of energy and a source of materials. Either of these can come from organic or inorganic sources. Members of the kingdom get their energy from fermentation of organic material, from photosynthesis, or from oxidizing inorganic substances.

The most important material that monera have to obtain from their environment is carbon. Some obtain carbon from organic material— these are the ones that are responsible for decay of dead plants and animals. Others obtain the carbon from inorganic compounds such as the carbon dioxide in the air.

98 **Monera can be aerobic or anaerobic.** Some monera are anaerobic, which means that they can carry out their energy generation only in the absence of oxygen. The bacteria that convert garbage piles into compost are examples of this class. Other monera need oxygen to survive, and are called aerobic.

I should point out that aerobic exercise has little to do with aerobic bacteria. Aerobic exercise is supposed to increase the body's use of oxygen, and is related to aerobic bacteria in name only.

99 **Perhaps the most unusual energy-creating mechanism among the monera is found in animals that live many thousand feet below the ocean,** near vents. The bacteria get their energy from the hydrogen sulfide (H_2S) that comes from the vents. These bacteria form the basis of a food chain that includes all sorts of crustaceans and giant worms.

100 **Species of single-celled organisms are not defined in terms of repro-** duction. I have to admit that I've always had a lot of trouble with biologists who talk about "species" of single-celled organisms. After all, members of the same species are supposed to be part of an interbreeding population. If there is no question of mating—if all reproduction is done by the splitting of cells—what does it mean to have a "species"? As it turns out, biologists use the term "species" in this case only in analogy to the way the term is used with more complex organisms. In practice, species of single-celled organisms are differentiated from each other by their biological niche, by the way they generate energy, and by the way the cell is put together.

101 **The most familiar "phylum" of the Monera are the bacteria.** The phylum bacteria is probably the branch of Monera that most people have heard of. Bacteria usually occur either as spheres (cocci), rods (bacillus) or in some kind of curved or corkscrew shape. Often they come together in colonies of many unrelated cells. Some biologists think that these kinds of colonies of cells may be the origin of multicellularity in higher life forms.

102 **Bacteria both cause and help cure human diseases.** We are familiar with bacteria because different species are responsible for a number of diseases in humans. Tuberculosis, strep throat, syphilis, dysentery, and cholera are all examples of diseases caused by bacteria. Lest you think that bacteria are an unalloyed plague, however, we should also point out that one species, *Streptomyces*, produces streptomycin, one of the most commonly produced antibiotics. In fact, we derive many antibiotics from bacteria. *E. coli* is a benign bacterium that has served as workhorse for the study of molecular biology.

103 Species of bacteria known as Chlamydia and Rickettsia are the smallest living things, being only a few hundred atoms across. They are, in fact, smaller than the largest virus. These species have about half as much DNA as do other species of bacteria. They are about as small as you can get and still be living.

104 Cyanobacteria account for a good deal of the oxygen and the photosynthesis that occur on the surface of the earth. This phylum of Monera includes single-celled organisms that float on the surface of water and are referred to loosely as "blue-green algae." It is believed that cyanobacteria were the first living things on the earth and that their waste product (oxygen) was partly responsible for the great change in the earth's atmosphere two billion years ago.

105 Cyanobacteria are one member of a group of organisms known as plankton. Members of this group are defined by the fact that they float on the surface of the water and do not swim. Plankton include bacteria and both single- and multicelled animals and plants. The term is contrasted to "nekton," or organisms that swim.

106 The kingdom of Protista is made up of single-celled organisms whose genetic material (DNA) is contained in a nucleus within the cell. It used to be that protists were classified as animals because they have a fairly complex structure, move around, and often surround and engulf their food. If you look at a drop of pond water under a microscope, all the little "animals" you see swimming around are protists. The most common protists move by means of flagella, long whiplike tails that the cell can move around to provide propulsion. Other protists move by means of cilia, tiny hairlike structures on the outside of the cell wall. The paramecium, which you probably encountered in your high school biology course, is an example of this kind of protist.

Protists are, in other words, quite a bit more advanced and more complex than the cells of the monera. They have a full complement of organelles and can be thought of as analogous in complexity to a large petroleum refinery. As one of my biologist friends told me, "Don't kid yourself—an amoeba is a very complicated thing!"

107 Protists make up a major part of the fossil record on earth. The protists known as Foraminifera are single-celled organisms that create a shell of hard material around them. They are very small (so small that they can only be seen with a microscope) but the shells are preserved in the sediment on ocean floors. Rocks like limestone are full of the fossils of the shell of this particular single-celled organism.

Examples of foraminifera.

Classical Genetics

108 **Genetics is the science concerned with the questions of how and why offspring resemble their parents.** Ever since human beings recognized the connection between sex and babies, it has been understood that there is a connection between the characteristics of parents and their offspring. The science of genetics is devoted to the study of what the relationships between parent and offspring are and how they come to be.

The next time you notice that little Willie has the same kind of eyes as Aunt Harriet, you are opening the door to one of the central fields of study in biology.

109 **Modern genetics was founded by Gregor Mendel.** The father (you should excuse the expression!) of modern genetics is generally considered to be an Austrian monk by the name of Gregor Mendel (1822–1884). Working in isolation in his hometown of Brno (in Czechoslovakia), he conducted a long series of experiments on pea plants that firmly established the basic laws of genetics. The content of his work (outlined below) is now usually called "classical" or "Mendelian" genetics.

Although Mendel did publish his findings in an obscure Austrian journal, his work remained largely unknown until after his death.

110 **Mendel's experiments were done with peas.** These experiments have entered the folklore of science and become an accepted part of our mythology. Working in his monastery garden, Mendel fertilized one set of selected pea plants with pollen from another set and observed the offspring. He quickly found that certain characteristics predominated in the resulting offspring. For example, if a tall plant was crossed with a short one, the result was not a medium-sized plant, but another tall one. If these hybrid offspring were crossed with each other (or allowed to fertilize themselves), one-fourth of the offspring were short, the rest tall. It was the discovery of these sorts of regularities that led Mendel to his theory of genetics.

A GENETICS GLOSSARY

111 **Allele**—the type of gene. For example, in Mendel's pea plants, one allele is for tallness, the other for shortness. The term refers to the genes themselves, not to the traits (such as tallness).

Genotype—a description of the kinds of genes (alleles) an organism has (as distinct from a description of the organism itself.)

Phenotype—a description of the characteristics of an organism. Thus, if I say one of Mendel's pea plants possesses a gene for shortness, I'm talking about its genotype, but if I say the plant is short, I'm talking about a phenotype.

---

112 **The basic unit of heredity is called the gene.** This is the term Mendel coined for whatever it is that is passed from parent to offspring. Your genes make you tall or short, give you blue or brown eyes, and so on. Today we understand that the gene is a string of many thousand molecules on a DNA chain, but Mendel didn't know what DNA was. He worked out his theory purely from observing his pea plants. In effect, the gene announced its presence to him through its effects on each generation of plants.

113 **The parents each have two complete sets of genes, and the offspring receives one gene for each trait from each parent.** The basic mechanics of the gene are simple. Each parent has two complete sets of genes, and each contributes one set of genes to the offspring. Which of the two genes in the offspring that is actually "expressed" (i.e., which characteristic is seen in the offspring) depends on what combination of genes is present, and these combinations are governed by the laws that Mendel discovered.

114 **Genes can be dominant or recessive.** When two genes come together in an offspring, there are well-determined rules as to which will go on to be expressed in the organism. If both the genes are the same—for example, if a child receives a gene for blue eyes from both parents—there is no problem. The child's eyes will be blue. But what if the child receives one gene for blue eyes, one for brown? The eyes can be only one color, so one gene must "win."

The gene that "wins" is said to be dominant. For example, in humans, the gene for dark eyes is dominant, so the child in the example will have brown eyes. The gene that "loses" is recessive. In Mendel's peas, the allele for tallness was dominant, while shortness was recessive.

The calculus of genes is very simple. If either (or both) genes are dominant, then the dominant trait is expressed. The recessive trait is expressed only if both genes are recessive. That's all there is to figuring out inheritance à la Mendel.

Take his peas as an example. In the first cross (tall and short), one parent had two dominant tall genes (which we'll abbreviate as T/T) while the other had two recessive short genes (s/s). Every offspring, therefore, had one tall and one short (T/s) gene and, since tallness is dominant, every offspring was tall. When the offspring were mated, however, on the average you would expect one T/T, one each of T/s and s/T, and one s/s. The first three will be tall, the last short, as Mendel found.

115 **You can carry a recessive gene and not know it.** The recessive gene remains in the DNA of the offspring and can be passed on to its descendents. For example, a child who has one brown-eyed and one blue-eyed parent may herself be brown-eyed, but carry a recessive gene for blue eyes. If she has a child with someone with a similar genotype, the child may be blue-eyed, even though both parents have brown eyes. A trait (such as blue eyes) that is carried but not expressed is said to be a recessive trait. So if Junior has blue eyes even though you and your spouse have brown, there's no need to suspect hanky-panky. This situation is well within the laws of Mendelian genetics.

In addition to blue eyes in humans, there are many other recessive traits. One of the best known is the gene for hemophilia, a condition in which the blood is unable to clot, so that minor cuts and contusions can cause death. This trait was common in some royal families in Europe, where inbreeding made it more probable that each seemingly healthy parent would carry hemophilia as a recessive trait that could be inherited by their children. Sickle-cell anemia, dwarfism, and Tay-Sachs disease are other examples of diseases and conditions related to recessive genes.

116 **In human beings, the following traits are recessive and dominant:**

Recessive	Dominant
blue eyes	brown eyes
color-blindness	color vision
baldness	hairiness

The gene for six fingers is dominant over the gene for five fingers as well. Odd, but true.

Pop Quiz

Someone who is color-blind marries someone who is not. The couple have four children. What is the maximum and minimum number of their children (on the average) who will be color-blind?

Minimum: 0 (if the dominant parent carries no recessive)

Maximum: 2 (if the dominant parent carries a recessive)

117 **Selective breeding is an example of Mendelian genetics.** Ranchers and farmers have known for a long time that it is possible to improve their livestock by selective breeding. For example, if your goal is to have cattle that grow fast and produce a lot of meat, then you allow only bulls that show these properties to breed. In this way, the genes that govern fast growth and meat production will be passed on to future generations according to the laws discovered by Mendel.

This application of Mendelian genetics explains two things that many people find puzzling: (1) why prize bulls can cost millions of dollars; and (2) how a breed like Black Angus—essentially a square chunk of beef on four short legs—could ever develop.

118 **Selective breeding inspired Darwin's theory of evolution.** In *The Origin of Species*, Charles Darwin devoted the entire first chapter to a discussion of what he calls "artificial selection," or selective breeding. His argument was that if people can cause such large changes in organisms by choosing which individuals have offspring, then nature should be able to do the same thing over longer times through natural selection. Although Darwin was unaware of Mendel's work, he unconsciously used Mendelian genetics in his arguments.

119 **The "green revolution" is a recent example of human use of Mendelian genetics.** In the 1960s, there was very real concern that the growing human population, particularly in the Third World, would outstrip food supplies. Plant geneticists, however, developed new strains of rice and other cereals that produced higher yields. The increased supply of food from land already under cultivation averted the looming catastrophe.

Classification of Living Things

120 **The great historical task of biology has been to find a way to order and classify living things.** The first thing you have to do if you want to understand the enormously complex and varied system of living things on our planet is to find some way of putting things in order—of sorting out which ones are close to each other, which are not. A typical question you might ask is "Is a human being more like a pine tree or a fish?" We owe our present scheme of classification to the Swedish naturalist Carolus Linnaeus (1701–1778). His scheme for organizing living things is a little like giving a house address by specifying country, state, city, zip code, street, and street number. In the same way, living things are located with successive degrees of precision by putting them in ever more restrictive categories until, at last, you come to a category that the living thing shares with no others.

121 It's not "animal, vegetable, or mineral" anymore. The old quiz game "Twenty Questions" always started with this information. The assumption in the game was that everything was either nonliving (mineral) or living (plant or animal). In this sort of broad classification of living things, "plant" and "animal" would be called "kingdoms." Today, biologists generally talk about five different kingdoms.

In addition to the traditional kingdoms of plants and animals, modern scientists recognize three others: Monera (single-celled organisms without a cell nucleus), Protista (single-celled organisms with cell nuclei) and Fungi (such as molds and mushrooms).

Specifying the kingdom for an organism corresponds roughly to specifying the country in giving an address.

122 The kingdoms correspond roughly to three levels of life. The monera and protists are single-celled organisms, with the monera representing the most primitive (and presumably the first to evolve). The three multicelled kingdoms (fungi, plants, and animals) each use a different strategy to survive. The fungi absorb what they need from the environment, the plants produce nutrients through photosynthesis, and the animals consume nutrients by eating plants, fungi, or each other.

123 Viruses do not fit into this classification scheme. All living things that are in the five kingdoms are made of cells. Viruses are not, since they contain only nucleic acids and protein. Are they living? That's hard to say, and is probably more a matter of semantics than anything else. In any case, they are ignored (or put into a footnote) in most discussions of classification.

124 Inside each kingdom, organisms are grouped into phyla. For example, in the animal kingdom there are dozens of phyla that take in worms, clams, insects, amphibians, reptiles, and many other "lower" life forms. Most familiar animals belong to the phylum of chordates—animals with a spinal chord, and to the subphylum of vertebrates—animals with backbones. Similarly, the plant kingdom has many phyla, including ferns, mosses, flowering plants, and so on.

Specifying the phylum or subphylum for an organism corresponds roughly to specifying the state when giving an address.

125 Although vertebrates are the most familiar animals, they are far from the

most common. Because we ourselves are vertebrates, we tend to think that this subphylum is much more important in the scheme of things than it actually is. In fact, most living things are not vertebrates—there are lots more beetles than mammals, for example. I always make this point to my classes by holding up an excellent book called *The Five Kingdoms* (by Lynn Margulis and Karlene V. Schwartz) that gives descriptions of all the phyla in all the kingdoms. The book is 374 pages long, with 4 pages devoted to chordates (animals with a spinal chord)!

126 **Phyla are further subdivided into class, orders, and families.** Continuing the subdivision into finer elements, we come to these three categories, which correspond roughly to giving the county, city, and zip code area in giving an address. Vertebrates, for example, are divided into classes such as mammals, amphibians, reptiles, birds, and so on. The class of mammals is further divided into several orders such as primates, rodents, carnivores, cetaceans (whales and porpoises), and so on. Finally, the order of primates is subdivided into families such as lemurs, various types of monkeys, great apes, and humans.

127 **Organisms are called by their genus and species.** The final subdivision in the Linnaean classification scheme is into genus and species. These correspond to the street and house number in an address. Organisms in the same genus (pl. *genera*) are similar to each other, but cannot interbreed. For example, the polar bear (*Ursus maritimus*) and grizzly bear (*Ursus horribilis*) are both members of the same genus in the family of bears, but are not the same species. Organisms that can interbreed with each other are members of the same species.

Biologists usually give only the genus and species when naming an organism, but the entire classification scheme outlined here is implicit in such a name. A familiar example of this usage: *Homo sapiens* (man the wise)—modern human beings belong to this genus and species.

128 **The breeding test for species doesn't always work.** Like most rules in biology, the "interbreeding criterion" doesn't always work. The timber wolf (*canis lupus*) can sometimes produce offspring with ordinary dogs (*canis familiaris*) even though they are classified as different species.

129 **Humans are the only living members of our genus and species.**
The human "pedigree" is as follows:

kingdom—animals
phylum—chordates
subphylum—vertebrates
class—mammals
order—primates
family—hominids
genus—homo
species—sapiens

130 Classifications are traditionally made by looking at anatomical similarities. Organisms are grouped together by using a series of increasingly restrictive criteria as we move down from kingdom to species. For example, mankind is first categorized as a member of the animal kingdom because, among other things, we are made of many cells (with nuclei) and ingest our food. We are vertebrates because we have a spinal cord encased in a backbone. We are mammals because we have hair and because our young are suckled. We are primates because, like monkeys and apes, we have big toes and thumbs, a large brain, eyes in the front of our head, and fingernails instead of claws.

When we come to separating the hominid family from other primates, the criteria become a bit more specialized. An example: the hominids (starting with *Australopithecus*) walked erect, rather than on their knuckles like gorillas. This means that the weight of hominid skulls rests in the spine, and fewer muscles are needed to balance it. This, in turn, means that hominids will not have attachments for heavy muscles on the back of their skulls, a fact that is used (among others) to differentiate them from apes.

Homo sapiens is distinguished from other hominids by even more detailed anatomical features, such as a flat face and larger sinuses around the eyes.

131 An organism can be extinct and still included in the classification scheme, provided, of course, that we have enough information about an organism's anatomy from fossils. This is why we can say that the dinosaurs were reptiles and *Australopithecus* was a hominid even though neither is around today.

132 There are competing schemes that some scientists want to use to replace the Linnaean classification. The Linnaean system is essentially static—we look at organisms as they exist today and group them together. We could look at evolutionary histories instead and, for example, group together those organisms that descended from a common ancestor. In this case, we would classify organisms as being together if there were few branchings on the family tree between them, and as far apart if there were many branchings. This approach is called cladistics and, for some reason that completely escapes me, seems to have become associated with Marxist political ideology.

A third, related, approach is to classify organisms by how long it's been since they shared a common ancestor and how much they've diverged since then. This is called phylogenetics. Both cladistics and phylogenetics concentrate on how organisms evolved to their present state, rather than on the details of what that state is.

133 The operational definition of a species remains a sore point in many fields of biology. The classical definition of a species is that two organisms are the same species if they can interbreed. Unfortunately, it isn't always possible to use this definition in practice. Take one example: scientists have identified millions of different species of beetles. Do you suppose they actually tried to breed each of these with all the others? Of course not—they simply looked at the anatomy of the insects and made a judgment based on their experience. Similarly, when species are extinct (like dinosaurs), you can't test the breeding compatibility, even in principle.

My favorite question when I want to annoy my paleontologist friends: if you had the fossils of a Chihuahua and a St. Bernard, would you identify them as the same species? About half of them say no.

quences of amino acids in the proteins the cell synthesizes, the overlap between proteins can be used instead. At the moment, the move to use this molecular measure of kinship is largely a gleam in the eye of its practitioners, primarily because the techniques for measuring overlap are so time-consuming and cumbersome. Nevertheless, DNA and protein-matching have been used in some instances, and it seems to me that it's only a matter of time until they are called on to replace the anatomy-based schemes of Linnaeus.

135 Individual humans share 99.8 percent of their DNA with other individuals, but "only" 98.4 percent with chimpanzees and 98.3 percent with gorillas.

Preview of Coming Attractions

134 In the future, relationships between organisms may be measured by similarities in DNA. A more direct measurement of the degree of relationship between two organisms is the amount of overlap between their genetic codes. Alternatively, since the DNA determines the se-

The Vest Button Award for Identifying a Species on the Least Evidence. There's an old saying that compares someone who leaps to conclusions to someone who "started with a button and sewed a vest on it." In the spirit of that saying we present the Vest Button Award to Canadian paleontologist Davidson Black who, in 1927, identified not only a new species of man, but an entire new genus (Peking man, or *Sinanthorpus pekinensis*) on the basis of a *single tooth!*

Plants

136 **Trace any food chain back far enough and you'll find a plant.** Plants supply the energy for all higher life forms on earth. Energy comes to our planet in the form of sunlight. Plants absorb part of this energy and, through the chemical reaction of photosynthesis, store it in the form of sugars, fats, oils, and starches. Herbivores eat plants to sustain themselves, and are themselves eaten by carnivores. In the process, the energy moves up the food chain.

137 **Most of the mass of living things is in plants.** We often fail to notice the plants that surround us—the grass on our lawns, the moss on a rock, the algae in a pond. Yet plants make up most of the mass of living things on the planet—at least 90 percent, according to most estimates.

Kinds of Plants

138 **Algae account for anywhere from 50 to 90 percent of photosynthesis on earth.** These are the simplest kinds of plants, and there are many kinds—anything from the single-celled organisms that float in water to large, highly structured organisms like kelp.

139 **Multicelled plants either have plumbing** or they don't. There are only two broad classifications of complex plants—those that move liquids around through an internal vascular system and those that don't. The first kind are called tracheophytes, the second bryophytes.

Mosses and some closely associated plants fall into the second category. They are the simplest plants that have enough structure so that they can support themselves on land. In this they differ from their ancestors, who were only able to float in water. Mosses also shelter their embryos after fertilization, rather than leave them to their own devices, as algae do.

140 **Most familiar plants are vascular.** By far the greatest number of familiar plants (grass, flowers, trees, etc.) are plants that have internal plumbing. This plumbing serves two purposes: it carries nutrients around the plant, and it provides the rigidity that the plant needs to stand up by itself. These plants also have leaves, which distinguishes them from the mosses.

There are two classes of vascular plants—the ferns and the seed plants. Ferns are the simplest vascular plants. They have leaves and plumbing, but they reproduce themselves with spores, rather than seeds. They were among the first plants to develop, but today they play a relatively minor role in the world ecosystem.

141 Complex plants reproduce with seeds.

Gymnosperms ("naked seed") are the simplest plants with seeds. Their seeds, once released, are not protected from the environment. The most common gymnosperms are evergreen trees, like pines and firs. The tallest plants in the world (redwoods) was well as the largest plants (Sequoias) are gymnosperms. Gymnosperms form the basis of our lumber industry and paper industries. The house you're sitting in and the page you're reading were probably made from gymnosperm fibers.

Flowering plants are the most complex plants, and they are also

THE VASCULAR PLANT

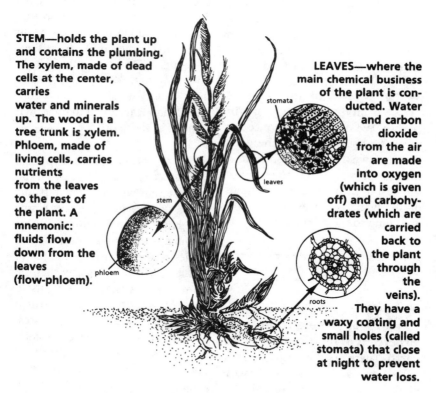

STEM—holds the plant up and contains the plumbing. The xylem, made of dead cells at the center, carries water and minerals up. The wood in a tree trunk is xylem. Phloem, made of living cells, carries nutrients from the leaves to the rest of the plant. A mnemonic: fluids flow down from the leaves (flow-phloem).

LEAVES—where the main chemical business of the plant is conducted. Water and carbon dioxide from the air are made into oxygen (which is given off) and carbohydrates (which are carried back to the plant through the veins). They have a waxy coating and small holes (called stomata) that close at night to prevent water loss.

ROOTS—can be fibrous (as in matted grass) or a taproot (as in an oak tree). Taproots allow the plant to get moisture even when the surface of the soil is dry. Roughly speaking, the roots of a tree spread out as much underground as the branches do above.

the most familiar. They are called angiosperms (from a joining of the Greek *angeion*, or vessel, and "sperm"). Their seeds are enclosed in fruit (which is actually an ovary), and the fruit are often adapted to give the seed widespread dispersal. Most of the food plants that we grow are angiosperms, as are ornamental flowers and hardwood trees.

142 **There are two important subclasses of flowering plants.** One, called monocots, includes grasses, lilies, orchids, and palm trees. Monocots have (among other characteristics) leaves with parallel veins and plumbing distributed throughout the stem (rather than in a layer near the surface). The other group, called dicots, includes trees, shrubs, herbs, and vines. They have all the plumbing in a ring surrounding the center of the stem, which is why you can kill a tree by cutting down to the wood in a continuous ring around the trunk.

Evolution of Plants

144 **Plants appeared on the earth over 3 billion years ago.** There is evidence in the fossil record that blue-green algae abounded in the earth's oceans 3.6 billion years ago. They were probably not much different from some simple types of modern algae. It is thought that the bryophytes (like mosses) and the vascular plants (like ferns and trees) evolved separately and at a different time from the primal algae.

145 **Flowering plants were once blamed for the extinction of the dinosaurs.** In fact, the appearance of flowering plants remains the great mystery of plant evolution. We know that they appear in the fossil record approximately 65 million years ago, but we do not have a very firm grasp on what their ancestors were or how they came into existence.

The dinosaur argument went like this: up until flowering plants came along, the dinosaurs ate things like pine needles, full of natural oils. When they had to switch to spinach, so the theory said, they all died of

---

TIMETABLE OF PLANT EVOLUTION

143 The "date" is the time this event would have occurred if we were to compress the history of the earth into a single year.

Time (million years)	Event	Date
3600	first algae	Mar. 21
433	land plants appear	Nov. 27
400	ferns and gymnosperms	Nov. 30
300 (approx.)	major coal deposits formed	Dec. 8
65	flowering plants appear	Dec. 26

constipation. (I'm not making this up.)

146 Sap flows in both xylem and phloem.

The term "sap," meaning the fluid you see when a plant is cut or damaged, refers both to what's in the xylem and in the phloem. The former fluid contains dissolved minerals, and the latter the end products of photosynthesis.

Pop Quiz

When we tap a maple tree for syrup, do we want the xylem or phloem sap? Answer: Phloem. We boil away everything except the sugars.

147 Unlike animals, which grow more or less uniformly all over, plants grow only at certain places.

These regions, where rapid cell division occurs, are called meristems. Some typical meristems on a plant are the tips of the root, the tips of the stem, and at lateral branching points (buds) on the stem.

148 The potato is actually a modified stem, not a root.

The eyes are lateral buds, so that when a potato sprouts, it's doing the same thing a tree does when it puts out branches.

149 Tree trunks grow just under the bark.

The region of cell division in a tree trunk, called the cambium, is the layer of slick stuff that surrounds the trunk just underneath the bark. In the cambium new xylem and phloem cells are made, expanding the trunk outward and leaving the older dead cells in the center. When a tree gets old, the tubes in the center get clogged with resin, producing what we call heartwood. This wood is highly prized because of its strength and its resistance to termites and rot. The part of the trunk where the tubes are still open, and therefore still carrying fluid, is called sapwood.

150 You can tell the age of a tree by counting its rings.

In the spring, the cambium produces large, thin-walled vessels. Later, when water starts to be less plentiful, the vessels get smaller and have thicker walls (to protect the water, which is now rarer and more valuable). The summer growth therefore looks darker. The alternation of light and dark bands is what we call tree rings.

Bristlecone pines, which grow in California, are the world's oldest living things. By counting tree rings in these pines, scientists can trace climate records back 8,000 years.

Pop Quiz

Why don't many tropical trees have annual rings? Answer: No seasons.

151
Dormancy is a survival mechanism for plants (even evergreens go into a state of reduced metabolism in the winter). Lower levels of sunshine in winter reduce photosynthesis, and by dropping their leaves, trees reduce water loss. In effect, during this period leaves become more trouble than they're worth.

Even during dormancy, severe cold can harm trees—if the water in the trunk freezes, it can expand and make the trunk explode. This causes a sound halfway between a loud snap and a rifle shot. No one who has ever heard that sound is likely to forget it.

152
One rotten apple really can spoil the barrel. The ripening of fruit and the falling of leaves are governed by the production of simple chemicals in the plant. For example, when the nights get longer, a tree's roots stop producing a chemical called cytokinin and leaves grow old and die. Eventually they fall off, wreaking havoc on suburban lawns in the process.

There is another chemical (ethanol) that seems to promote the falling of leaves. This was discovered in the late nineteenth century when it was observed that trees near gas lamps lost their leaves earlier than elsewhere. Today, tomatoes are often picked green, then given a whiff of ethanol to turn them red en route to the supermarket. This is why winter tomatoes don't taste like the ones from your garden.

A rotting apple gives off ethanol, making other apples in the barrel ripen quickly. This is the chemical origin of that particular piece of folk wisdom.

153
Leaves don't actually "turn colors" in the fall. Leaves normally look green because they contain chlorophyll. When the leaf dies the chlorophyll disappears and the other colors, associated with substances that were there all along, emerge.

Ecosystems

154
An ecosystem is the totality of plants and animals in a given region, together with the physical environment in which they live. The science devoted to the study of ecosystems is called ecology.

155
In ecosystems, atoms cycle and energy

flows. The atoms that make up the world ecosystems today are the same atoms that made it up a million years ago and will make it up in the future. Atoms move through one part, then another, but never leave. In smaller ecosystems (such as a lake), atoms may still cycle, but the strictures are less severe and there can be a flow of material in and out across the boundaries. An ecosystem which does not exchange material with its surroundings is called a closed ecosystem; one which does is called open.

Virtually all the energy in the world ecosystem comes from the sun, stays for a while on the planet and is then radiated back into space in the form of infrared radiation. Thus, the energy does not stay on earth but simply passes through on its journey from the sun to outer space. This behavior is typical of the way that energy interacts with ecosystems.

156 **Carbon is cycled through the global ecosystem.** Carbon is take from the atmospheric supply of carbon dioxide by photosynthesis and incorporated into the tissue of plants. Animals eat the plants (or other animals that have eaten plants) and the carbon is taken into their bodies. The respiration of these animals returns a portion of carbon to the atmosphere in the form of carbon dioxide, thereby completing one loop of the cycle.

If the plant is not eaten, the original complement of carbon will still be in its tissues when it dies. This carbon can enter into long-term storage in the form of fossil fuels like coal and petroleum. It can decay and be digested by bacteria, in which case the carbon will be returned to the atmosphere eventually in the form of CO_2. The carbon in the body of animals is returned to the atmosphere in a similar way once the animal dies.

A major "sink" of carbon is carbon dioxide dissolved in the deep waters of the ocean. There is much more carbon stored in the ocean than exists in the atmosphere.

If fossil fuels are brought to the surface and burned (as they are when we generate power with coal or drive automobiles), then their complement of carbon is taken out of storage and returned to the atmosphere.

157 **Nitrogen is cycled through the world ecosystem.** Although there's a great deal of nitrogen in the atmosphere, most living systems cannot use it directly. Nitrogen enters living things through the action of nitrogen-fixing bacteria. Once nitrogen is incorporated into plant tissue, it can move into the tissue of animals that eat plants. Nitrogen is returned to the soil when plants die and in the wastes of animals, and bacteria can return nitrogen from the soil to the atmosphere. Like carbon dioxide, nitrogen is stored in great quantities in the form of dissolved gas in the ocean.

158 **You can't throw anything away,** since all

materials must cycle through the world's ecosystem. No matter how deeply something is buried or how far out in the ocean it is dumped, it remains in the ecosystem and will eventually come back, perhaps to haunt you. This is one of the great truths that governs modern policy debates on questions of pollution.

159 Energy flows up the food chain. In any ecosystem, the food chain expresses the relation between living organisms, indicating which are eaten and which do the eating at each step. At the lowest level in the food chain are the plants, which create their own tissues directly from sunlight. We can call plants the first trophic level (a trophic level is a group of organisms that get their energy the same way).

Animals that eat plants (herbivores) form the second trophic level. It is obvious that herbivores, like plants, do not make efficient use of their primary energy input. Most plants die a natural death, unconsumed by rabbits and their kin. Typically, herbivores might be able to utilize 10 percent of the energy available in the first trophic level.

A third trophic level consists of primary carnivores (like wolves) that eat herbivores, and a fourth of animals (like killer whales) that eat primary carnivores. Finally, at the end of the food chain, there are animals (like vultures and some insects) that consume dead plants and animals.

160 Human beings, grizzly bears and other omnivores feed from all trophic levels. They are probably the most efficient of the animals in terms of utilizing energy that comes up to them through the food chain.

161 Once you understand that moving from one trophic level to the next involves a loss of about 90 percent of the available energy, you can understand a number of facts about food prices. If you lose a factor of ten in energy each time you move up one trophic level, you will have to pay ten times more for the energy in the higher trophic level than you would for energy in the lower trophic level. Thus, for example, beef costs about ten times as much as grain by weight because the energy in the beef has to be obtained from the grain at a loss of about a factor of ten in energy.

162 Pollutants are concentrated as they move through the food chain. Just as energy becomes more concentrated as we go from plant to herbivore to carnivore, so do any pollutants that have entered the food chain. This is a cause for concern among ecologists and public policy analysts.

Pop Quiz

Why don't we raise lions for food? Answer: Lions are carnivores in the third trophic level, so, ignoring shipping costs, lion meat should cost about ten times as much as beef—far too expensive for any but the most dedicated gourmet.

Populations

163 **Populations grow exponentially unless checked.** In exponential growth the number of offspring in any generation is proportional to the number of individuals in the previous generation. For example, if each individual in a population produces two offspring which survive to maturity, that population will grow exponentially.

164 **The most important concept relating to exponential growth is that of doubling time** —the time it takes for a population to double. A rough formula for calculating doubling time is:

$$\text{Doubling time} = \frac{70}{\%\ \text{annual increase}}$$

Thus, a population which grows at 10 percent a year will double in numbers in seven years.

Pop Quiz

With a 5 percent rate of inflation, how long will it take for your dollar to be worth half of what it is today?

Answer: $70 \div 5 = 14$

Fourteen years from now that dollar in your pocket will buy what 50 cents buys today.

165 **Exponential growth cannot continue indefinitely.** Sooner or later something has to give. In nature populations grow exponentially until they run out of food or until predators begin controlling their numbers, at which point the population levels off. A typical growth in numbers of population is shown.

One of the attributes of the human population that worries demographers is that it has a doubling time of about thirty years, so that the number of humans in the world will double by 2020 unless measures are taken to reduce birth rates.

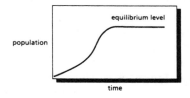

The growth of a population starts out being exponential and then, as it approaches the carrying capacity, levels off.

166 **Sooner or later even the largest resource base will be depleted by a population that continues to grow.** Ecologists refer to the number of living things that can be supported by a given ecosystem as its "carrying capacity." As a population approaches the carrying capacity, it must level off and become constant.

167 **Populations of one type of organism can be limited by other organisms.** When more than one type of organism is present in an area, they can interact by competing, by predation, or they can actually benefit one another in some way. An example of the first type of interaction would be two species of gazelles grazing the same plain. They compete because grass eaten by one cannot be eaten by the other. An example of the second type of interaction is seen in the population of rabbits and coyotes in a given area. The rabbits would overpopulate if it weren't for the coyotes, but on the other hand the coyotes could not exist without the rabbits. Each population acts as a check on the other.

Mutually beneficial relations between species is somewhat rare in nature. One example is the relation between flowering plants and honey bees.

168 **In general, if predators are removed from an ecosystem, the number of prey species will grow without bounds until it reaches the carrying capacity,** at which time there is widespread famine. Conversely, the number of predators cannot grow too large without depleting the prey species and, in the end, the number of predator species as well.

There are many examples in the United States of deer populations exploding and becoming subject to starvation and disease when hunting by humans is limited. In many cases, human hunters play the role of the now-vanished predators like the timber wolf and mountain lion. Although I am not a sport hunter, I recognize the important role that hunters play in controlling the numbers of some species in the American wild. It's sometimes hard to remember that we, too, are part of the natural system.

2

EVOLUTION

The English peppered moth (see item 171) is living proof of the power of natural selection.

Evolution

169 **Life on earth developed through the process of evolution.** This statement includes everything from bacteria to pine trees to giraffes. The concept of the evolution of life provides the central framework around which the life sciences are organized. Because all these fields share an evolutionary view of life, it is possible for someone studying the ecosystem of a large lake to talk the same language as a colleague studying the sequence of molecules along a particular piece of DNA, even though it might appear at first as if they have nothing in common. It is impossible to understand the modern biological sciences without understanding evolution.

170 **The central mechanism of evolution is natural selection.** The basic idea of natural selection is this: at any given time there are variations in a population. Some giraffes have longer necks than others, some human beings can run faster than others, and so on. If certain variations give individuals who possess them a higher probability of surviving long enough to have children, then those characteristics are more likely to be passed on to the next generation. For example, if having a longer neck allows a particular giraffe to eat leaves that other giraffes cannot reach during a drought, the giraffe with the long neck is more likely to survive the drought and have offspring. These offspring will resemble their parents and have longer necks. If the long neck remains advantageous, then over long periods of time giraffes with long necks will eventually become the dominant variation in the population. In this way, a characteristic that allows an individual to exploit his environment more efficiently comes to be shared by all members of that species. This is basically all there is to the idea of natural selection.

171 **Evolution is still going on today.** The development of life is not a process that took place at one time and then stopped—living things still adapt to their environments today. The most famous historical example of this is the history of a type of moth that lived in central England. Originally these moths were mottled white and brown so that they blended in with the trees that were their natural environment. During the Industrial Revolution, the English environment became much darker because of the soot from the factories. In response to this, the moths evolved in a few short years to a grayish color to blend with the new environment. When the clean air movement hit England in the 1960s and the factories were cleaned up, the moths started to evolve back to their original colors.

172 Charles Darwin (1809–1882) is the founder of modern evolutionary theory. Shortly after finishing his university studies, Darwin signed on as a naturalist for a five-year voyage of exploration on a ship called the *Beagle*. During this trip, he became convinced that species were not immutable, but changed gradually

Charles Darwin.

over time. One study that led to this conclusion was that of finches on the Galapagos Islands, where closely related birds on different islands had developed quite different characteristics (beak shapes, for example) in response to their different environments.

In 1859, Darwin published *On the Origin of Species by Means of Natural Selection*, perhaps one of the most influential books ever written. Although strongly opposed by theologians at the time, Darwin's view of life has since become accepted by all but a fringe element among religious thinkers. There is so much evidence to support it that scientists scarcely bother to think about its basic validity any more, but concentrate on working out the fine points of the theory instead.

In what must rank as one of the most incorrect assessments of a young man's talents ever made, Darwin's father responded to his rather poor scholastic achievements by writing "You care for nothing but shooting, dogs, and rat-catching, and you will be a disgrace to yourself and your family."

173 There is a difference between the fact of evolution and the theory of evolution. You sometimes hear people comment that evolution is "just a theory." This statement is very misleading, because evolution is a fact as well as a theory. You can understand what I mean by thinking about gravity. There are theories of gravity, including those of Newton and Einstein. These theories can be right or wrong, complete or incomplete. But there is also the *fact* of gravity—when you drop something, it falls. This fact remains whether the theories are right or not.

In just the same way, the record of the development of life from humble beginnings to the present complex biosphere is a fact that can be read in the fossil record. Similarly, the ability of living things to mutate in response to environmental change can be seen in laboratories and in nature. Whether any of the current theories of evolution can or cannot explain this record doesn't change the existence of the fact of evolution.

This is an important distinction, because one of the favorite techniques of creationists is to argue that since scientists are in disagreement about this or that point of evolutionary *theory*, the *fact* of evolution must be thrown out and their version of creation accepted. This is roughly equivalent to overhearing two people arguing about whether an office is on the 52nd or 53rd floor of the Empire State Building and concluding that the building must be one story high.

The panda's thumb is one of the best pieces of evidence for evolution.

out that well-adapted organs like the eye cannot be used to prove the theory of evolution, since these organs could equally well be explained in terms of a special creation. Organs like the human appendix or the panda's thumb, however, do provide such evidence.

The panda, a distant relative of the raccoon, lost its true thumb early in its evolutionary history. When the environment in which it found itself changed and bamboo leaves became a staple of its diet, there was an advantage to be gained by having something like a thumb to strip leaves from stalks. In the panda, a thumblike spur evolved from the side of a bone in the wrist.

This is, of course, a very awkward way to give the panda a thumb—not at all what you'd do if you were designing the animal from scratch. The mechanism of a natural selection takes an animal from what it is and adapts it to whatever the environment happens to be. It doesn't necessarily produce the best or even the most efficient organism possible—just the best that can be made from the material at hand. Sometimes, as in the case of the panda, the result has a decidedly jury-rigged appearance.

Enduring Mystery

174 Nature's mistakes may be the most dramatic evidence for evolution. In his marvelous essay "The Panda's Thumb," Steven Jay Gould points

175 How did wings evolve? While the evolutionary advantage to be gained from fully developed wings is not hard to see, the same can't be said about the advantage to be gained from the rudimentary appendages that must have led to them. In some cases, for example, birds, the wings evolved from arms and hands. For insects, however, the wings must have evolved from protuberances on the animal's side. Why would such protuberances have conveyed any advantage? The fact that wings would help a descendant a million years in the future certainly couldn't help an individual survive today.

Recently, scientists have argued that the protuberances played a role in temperature regulation—they provided extra surfaces through which heat could be absorbed or radiated. Calculations show that the most efficient heat exchangers are just about big enough to allow the insect to glide (something like a modern "flying" squirrel). From that point on, the organ, originally developed for one purpose (heat transfer), could be used as the basis for the development of another (flying). This idea, which makes a lot of sense to me, illustrates the ad hoc nature of the evolutionary process very well.

176 Evolution of life on earth proceeded in two stages: chemical and biological. Life on earth must have developed from inorganic materials—what else was there for it to come

---

IMPORTANT DATES IN EVOLUTION

177 The following table summarizes important events in the development of the earth. The "date" in the right-hand column refers to the day the event would have occurred if we compressed the entire history of the earth into a single year.

Time (in millions of years ago)	Event	Date
4600	earth forms	Jan. 1
3800	oldest known rocks form	March 5
3600	oldest known fossils (algae)	March 21
2000 (approx.)	significant oxygen in atmosphere	July 26
650 (approx.)	multicelled life in oceans	Nov. 10
590	fossil record begins	Nov. 14
440	life moves to land	Nov. 25
400	fish (vertebrates) abound	Nov. 29
250	dinosaurs appear	Dec. 12
65	dinosaurs become extinct	Dec. 26
*****From now on, all dates refer to times on Dec. 31*****		
4	first hominids	7:30 AM
0.1	first *Homo sapiens*	11:49 PM
0.005	recorded history	11:59:34 PM

from? The first stage in the development of life, therefore, was the production of a reproducing cell from materials at hand on the early earth. This process is called chemical evolution (see item #187 and following). Once a living, reproducing system was present, the process of natural selection took over to produce the wide variety of life that exists today.

Continuing Questions in Evolution

178 **How did evolution take place?** When Charles Darwin first proposed the theory of evolution, he thought that changes in organisms took place a little at a time, with each generation's changes building on those of the last until the accumulation produced the dramatic shifts we see in the fossil record. This idea has come to be known as gradualism. In 1972, two American paleontologists, Steven Jay Gould and Niles Eldrige, proposed an alternate view of evolution. Their interpretation of the fossil record was that throughout most of the past little change occurred from one generation to the next, a phenomenon they called "stasis." These periods of tranquillity, in their view, were punctuated by short bursts of rapid change. This interpretation of the fossil record is called "punctuated equilibrium."

The debate over which of these two interpretations of the fossil record is correct continues because the fossil record is so spotty and incomplete that we can't really tell the difference between them. For the record, I think the answer to the question "How did evolution take place?" is "All of the above." There are probably examples of both rapid and gradual change in the history of life. The world is too complex to have simple answers!

179 **Did life really start on other planets?** The conditions that have to be met for life to evolve from inorganic matter on earth are pretty stringent. The seeming improbability of life developing here has led some people to suggest that life came to our planet from elsewhere. This notion has come to be known as "Panspermia." In the nineteenth century it was suggested that life was carried from one star system to another by some kind of spore, but this notion fell into disrepute when it was realized that the radiation such a spore would encounter in deep space

One view of the way life may have developed on other planets.

would far exceed any reasonable lethal dose.

More recently, a variant of this notion, called "Directed Panspermia" was put forward by Nobel laureate Francis Crick. His idea is that extraterrestrial civilizations put microbes into shielded spacecraft and sent them to seed likely planets. The main problem with this view: how did life develop on the progenitor's home planet? After all, the things that make life unlikely on earth also make it unlikely everywhere else. Why substitute two miracles (life plus the desire to seed the universe) for one (life alone)?

Can Darwin's ideas be applied to societies? One of the most interesting (and controversial) extensions of Darwinian ideas is the study of sociobiology. The essential premise of sociobiology is that some principles of biological evolution hold for the evolution of cultures as well as organisms.

My reading of the sociobiology situation is this: the theory has weathered an initial fervent opposition, based primarily on ideology and centered in the political left. Sociobiology is now in a period of intense development which will end when we find out how far this idea can be carried to explain human social structure.

It's hard to imagine anyone less likely to be at the center of controversy than Edmund O. Wilson of Harvard University. He is a quiet, scholarly man whose first love is the study of ants, both living and extinct. He has, for example, a long-standing arrangement with Haitian amber dealers to give him first refusal on old ants trapped in amber.

But Wilson's work in the evolution of insects eventually brought him to his present status as founder and guiding light of the new science of sociobiology. As such he has been vilified by his colleagues, criticized in the press, and shouted off the platform at scientific meetings by radical students. It has taken, I think, more than a little courage for him to continue to follow his ideas in the face of this outcry.

Common Misconceptions About Evolution

180 Evolution does not say that humans descended from apes. It is an old canard, going back to the days of Darwin, that evolution requires that human beings descend from apes. In fact, the theory teaches that both humans and apes descended from a common ancestor that lived millions of years ago.

181 Evolution does not require a "missing link" between humans and apes. The search for the "missing link" between humans and apes occupies a special place in popular mythology. My own favorite example of the power of this image is a professional wrestler who painted his face green and called himself "The Missing Link." The idea behind the image is that if mankind descended from apes, there should be a creature that is half ape, half man. In fact, since mankind and the apes descended from a common ancestor, there should be no such thing.

182 **"Survival of the Fittest" doesn't mean what it seems to.** Natural selection is often characterized as "the survival of the fittest." Darwin himself used this phrase, but it is often misrepresented or misunderstood. Darwin used the term "fit" to describe individuals who are successful in producing offspring in the next generation, nothing more. In general, those individuals that are best adapted to their environment will be "fit" in this sense.

During the nineteenth century and persisting into our own times the notion of "fitness" was given a moral overtone by many philosophers. It was argued that the "best" survive and prosper. As you see by the example of the giraffe this is not the way natural selection works. There is no moral judgment in nature. The only statement you can make is that those individuals whose genetic makeup gives them an edge in competition with their fellows are more likely to have offspring, so that at some distant time in the future those offspring will dominate the population.

During the nineteenth century the British philosopher Herbert Spencer adapted what he thought were Darwin's ideas to social criticism. His ideas have since come to be known as Social Darwinism. The basic idea of Social Darwinism is that just as nature operates according to the survival of the fittest, so does society. The rich, according to Spencer, got where they are because they are fit while the poor are where they are because they were not fit.

Of course, this stands the entire Darwinian paradigm on its head. In fact, someone like Leland Stanford, the builder of the Southern Pacific Railroad and one of the great robber barons of nineteenth-century America, was decidedly unfit by Darwinian standards. He had only one child, a young man who died before he could sire children. On the other hand, the lowliest Chinese coolie or Irish laborer on Stanford's railroads might easily have had a dozen children and was therefore, in the Darwinian sense, much more fit than Stanford.

When I'm discussing the Darwinian theory with classes of undergraduates, I delight in pointing out to them that by going to college they are making themselves unfit because they are squandering their best reproductive years.

183 **Individual members of a species cannot change their genetic makeup.** The French scientist Jean-Baptiste Lamarck believed that acquired characteristics could be passed on from one generation to the next. For example, if the giraffe stretched to get at leaves, his neck would grow longer and his children would inherit that acquired characteristic. Today we know that such things are not inherited. A weight lifter's child does not automatically get big muscles, nor does a marathon champion's child automatically get increased lung capacity. We inherit a great deal from our parents, but this inheritance has nothing to do with the strivings of our parents.

During the 1920s, the Russian "geneticist" Trofim Lysenko achieved political dominance in the Soviet Union because his theories were thought to be consonant with Marxist philosophy. He rejected the idea that genes had anything to do with inheritance, and turned his back on "decadent" Western science. He promised Stalin that he would grow a line of lemon trees from the Black Sea to Moscow by planting them in successively colder climates and allowing each lemon tree to adapt to its climate before planting its descendants a little farther north.

Using his political influence, Lysenko crippled Soviet biological science for half a century by sending his competitors to the Gulag and by outlawing the teaching of modern genetics. The Lysenko affair remains one of the blackest episodes in the history of science.

Creationism and Evolution

184 **Creationism is the doctrine that the earth was created a few thousand years ago by a divine act.** Creationism, or "creation science," has enjoyed a modest revival in the United States. It is the doctrine that the earth was created a few thousand years ago, more or less as described in the Book of Genesis. It holds that living systems were specially created in their present form, and that no changes have occurred since the creation itself. In general, creationism is associated with conservative Protestant churches in America, and has little support from either mainline science or mainline theology.

185 **Creation science does not operate like a science.** Creationists have tried to argue that their views ought to be given equal weight in the public schools with the teachings of the accepted theory of evolution because they represent an alternate "science." Fortunately the courts have held that this tactic is simply a roundabout way of introducing religious teaching into the public schools. From a scientific point of view the problem with creationism is that it is not possible to prove it wrong. No matter what evidence you find, the answer is always, "Well, that's the way things were made."

For example, one common argument against the creationists is that we can see stars that are billions of light-years away, so light must have been traveling toward us for billions of years. Therefore, the earth could not have been created six thousand years ago. The creationists' answer to this is what they call the doctrine of created antiquity. In essence, they argue that the light was created on its way here in just such a way that it would look like things were billions of years old. I have to admit that I find the prospect of God as the ultimate practical joker a little hard to swallow.

186 **Evolution does not violate the second**

law of thermodynamics. As a physicist, there is one argument from creationists that really sets my teeth on edge. This is the argument that evolution requires that life go from simple to complex, whereas the Second Law of Thermodynamics says that systems move to a state of maximum disorder, and that, because of this, evolution violates the laws of physics.

The problem with the argument is that the Second Law holds only for isolated systems. The earth is not such a system because it is constantly receiving energy from the sun. To see why this detail is important, you can think about a common activity—making ice cubes in a refrigerator. When you make an ice cube, you create a system of high order (the ice) from a system of low order (the water) by using energy from your local utility. The increase in order in the ice cube is balanced by greater disorder at the generating plant, where the burning coal heats the atmosphere. So long as the overall books balance, there is no violation of the laws of physics.

The same argument works for living systems on the earth. The increased order in the biosphere is balanced by increased disorder in our "power plant"—in the sun.

After all, if the creationist argument were right and it was impossible for *any* system to become more ordered, you could never make ice cubes to cool your drinks!

Evolution of Complex Life

Chemical Evolution

187 We understand how the basic building blocks of life could have evolved on the early earth. In 1955 Harold Urey and Stanley Miller at the University of Chicago performed an experiment that showed how the first step in chemical evolution may have taken place. They put together methane, hydrogen, ammonia, and carbon dioxide—the materials we believe were the components of the earth's early atmosphere—and subjected the mixture to electrical sparks (thus simulating the effects of lightning). In a matter of hours they noticed that the constituents of early earth were forming molecules known as amino acids, which are the basic building blocks of proteins. Proteins, in turn, are the molecules that carry out most of the chemical work in living systems. What Miller and Urey had done, in other words, was to start with nonliving materials and produce the simplest materials that make up the living cell.

Subsequent experiments have

shown that not only lightning, but heat (for example, from volcanoes) and ultraviolet radiation (for example, from the sun) can produce amino acids from the same materials. In fact, modern researchers have found that Miller-Urey type reactions can be used to create not only amino acids, but a whole alphabet soup of biochemical molecules.

188 Life might have started in the primeval soup. If the Miller-Urey reactions went on in the atmosphere of the early earth, there would have been a rain of amino acids on the ocean. In about one hundred thousand years (which is a short time, geologically speaking), the ocean would have had the same concentration of amino acids as it now has salt. Thus, the ocean would have been teeming with molecules which could be put together to make living systems.

The ocean that resulted from this rain is often called the primordial soup. It had a concentration of amino acids of several percent—roughly the same as you'd get by putting a cube of bouillon into a gallon of water. It wouldn't have been a pleasant place to swim—a lot of the amino acids are pretty smelly—but it was a place that was extraordinarily rich in nutrients. It is in this soup that we believe the first living cell developed.

189 Some amino acids may have been brought in by meteorites. One of the most surprising developments of the last few decades has been the discovery that amino acids are quite common in the universe. You can see them in giant clouds out in the Milky Way and on asteroids that fall to earth from space. This has led some people to suggest that some or all of the amino acids in the primeval soup came to earth with meteorites. Whether meteorites, the Miller-Urey process, or both contributed to the buildup, the result was that, shortly after the earth cooled, the oceans were rich in amino acids.

Enduring Mystery

190 We do not know how the first living cells formed from the complex molecules in the primordial soup. This is the greatest gap in our knowledge of the evolution of life.

The problem with a cell developing in the primordial soup is a classic catch-22 situation. If amino acids come together to form proteins in the air or at the surface of the ocean, the ultraviolet radiation from the sun will break the proteins up. To escape this fate, the amino acids have to combine underwater. But if they do so, chemical reactions with the water will break them up. The only way that a soup of amino acids could develop into something more complex would be for the concentration of amino acids to be so high that there were regions where the water couldn't get in and break up the more complex molecules. Because of this, conventional thinking about the first cell focuses on ways that amino acids could have become concentrated.

This concentration could take place in tidal pools, where water would slosh into the pool at high tide and then evaporate during low tide, leaving the amino acids behind. If the pool was as much as 30 feet deep, then the ultraviolet radiation at the bottom would be sufficiently screened so that combinations of amino acids could stay together at the bottom.

Alternatively, heat from a volcano could evaporate water and allow the amino acids to concentrate. A third possibility is that the chemical combination of the amino acids can take place between the layers of some clay minerals at the ocean bottom.

191 **Although we do not know how the first cell developed, we do know that it happened very quickly.** We date the formation of the earth at 4.6 billion years ago, and fairly advanced single-cell organisms were in existence about 3.6 billion years ago (see below). This means that we have at most 800 million years in which to go from a hot, totally inorganic planet to a cool, living one. In fact, the current best guess is that less than a million years after the ocean formed, it was full of single-cell living systems.

192 **The first cell may also have formed in a primordial oil slick.** Some fatty materials form spontaneously into bubbles when they are put in water. This fact is used to formulate another scenario for the formation of the first cell. In this scheme the early ocean has some of these fatty materials in it—sort of a primordial oil slick. They form bubbles and the chemical reactions that allow amino acids to build up into proteins take place inside those bubbles, protected from the water. This scenario has the advantage of solving two of the major problems encountered in forming the first cell: it explains how complex molecules are able to form, and it also explains how the contents of the cell were isolated from the environment once they did.

The Fossil Record

193 **We know about the evolution of life through reading the fossil record.** When a plant or animal dies, it may be buried. Underground water flows around the remains, and gradually minerals in the water replace the atoms in the buried organism. The end result, after a long period of time, is an exact replica in stone of the buried parts. Millions of years later, a paleontologist finds the fossil and a new piece of information about past life is added to our store. The total information contained in the fossils that we have discovered is called the fossil record.

194 **The fossil record is imperfect.** Not every animal that dies becomes a fossil—it is much more usual for carcasses to decompose without ever being buried. Not every plant and animal that gets buried becomes a fossil—obviously hard parts (like skeletons) are going to be preserved more easily than soft parts like skin and organs. And, finally, not every fossil that ever was created has been discovered. Paleontologists estimate that about one species (*not* one animal or plant) in ten thousand is known to us through the fossil record.

For all its imperfections, the fossil record is the only game in town when we want to learn about how life developed on our planet.

195 Most of the fossils we have come from continental shelves.

When a plant or animal dies on land, its skeleton is likely to be scattered by scavengers and the weather—its chance of making it into the fossil record is small. Organisms on the continental shelf, however, fall into the rich ooze at the bottom and are buried immediately, greatly increasing their chance of becoming fossilized. It should come as no surprise, then, that the great preponderance of fossils we have are those on plants and animals from areas that were continental shelves.

This isn't as bad a situation as it seems at first—after all, if you had to pick one type of region to evaluate the health of the earth's present ecosystem, you'd probably pick the continental shelves anyway.

196 The first cells that left a mark were blue-green algae that lived near the side of the ocean.

You can think of them as being like the green scum that often collects near the edge of ponds and lakes and in stagnant reaches of a river. The algae themselves did not leave fossils, but mats of algae left their imprints on clays and muds that later changed into rocks. These layers of mud and clay, together with their imprints, eventually formed into rocks, and these rocks can be seen in a few places at the surface of the earth today.

197 We may carry the imprint of the first cell in our own bodies.

The DNA in our bodies and all living things is in the form of a right-handed spiral. Why should all living things on earth share this particular kind of

Stromatolites forming in Australia today. The first fossils may have looked like this.

DNA? After all, a DNA molecule can equally well appear in a left-handed spiral.

One particularly intriguing suggestion is that the first cell, when it developed, happened to incorporate right-handed DNA into its structure. There's no reason why it had to do so, it's just that it had to be either right-handed or left-handed. However, once the first cell developed, it very quickly multiplied in the nutrient-rich primordial soup. It probably would have taken a very short time for this particular kind of cell to have taken over the earth and used up all of the simplest and most easily utilized nutrients available. If another cell (perhaps having left-handed DNA) developed someplace later, it would find itself in competition with a species which had already been acted upon by natural selection and been made very efficient. The second cell wouldn't have a chance. Consequently, the argument goes, your right-handed DNA may well be a vestige of that first cell that developed somewhere in that old ocean.

Highlights of Evolution

198 **The "real" fossil record begins about 600 million years ago.** Most of us don't think of things like mats left by algae when we hear the word "fossil." Instead, the word conjures up images of huge dinosaur skeletons in dimly lit museum halls.

It wasn't until about 590 million years ago, at the beginning of what geologists call the Cambrian era, that living things began to develop skeletons and hard parts. These parts, because they tend to last longer than ordinary tissue, are much more likely to become fossils. Until recently, the fossil record looked pretty barren until the Cambrian, but there was an abundance of fossils thereafter. Scientists often spoke of this sudden appearance of fossils as the "Cambrian explosion."

Today, we realize that complex life existed before the Cambrian, but that it left a very sparse fossil record—think of the oceans as having been full of things like jellyfish. What "exploded" in the Cambrian, then, was the population of organisms with skeletons.

199 **Complex life began in the ocean.** Just as the first cell probably developed in the ocean, so too did the first organism with more than one cell. Five hundred ninety million years ago a complex assemblage of plants and animals developed in shallow ocean waters—things like seaweed, clams, corals, and so on. For over 150 million years all life on earth was in the sea and the land was barren.

200 **Life moved to land about 430 million years ago.** Plants moved to land first, then animals like present-day scorpions. This happened during what geologists call the Silurian period. Because there were no competitors in the new territory, organisms that moved to land did very well and spread quickly.

201 Fish were the first vertebrates. Sharks and huge armored fish (now mostly extinct) dominated the oceans about 380 million years ago, during the Devonian era. At that time, they were the most complex life form on the planet.

202 The "Age of Reptiles" started 248 million years ago and ended 65 million years ago with the extinction of the dinosaurs. This is the period most familiar to people because it includes the dinosaurs. Dinosaurs were a diverse family, and they weren't all huge. Many were no bigger than a modern fox. At the same time that dinosaurs dominated the land and sea, ancestors of modern mammals in the form of small, mouselike creatures, eked out a precarious existence.

203 The extinction of the dinosaurs cleared the way for mammals. For the last 65 million years, mammals have been the dominant life form on earth. Some people have argued that whatever drove the reptiles to extinction was a good thing—otherwise reptiles would have continued to run things and humanity would never have evolved.

204 Humans have been around for a very short time. If you take the age of the earth to be one year, then "humans" have been around for only the last few hours. That's if we count our ancestor Lucy as human. *Homo* *sapiens* himself has been around only a few minutes.

Dinosaurs

205 From the point of view of science, dinosaurs don't count. Well, OK—they count some, but not nearly as much as most people think. There were never more than a few species of large dinosaurs alive at any given time. They were fascinating—who can forget *Tyrannosaurus rex, Brontosaurus,* and *Tricerotops?* Nevertheless, they were analogous to modern elephants and rhinos: beautiful and interesting, but carrying little information about life on earth in general. Add to this the fact that dinosaurs, because they were land animals, rarely left fossils, and you have a situation where the kind of fossil that is most interesting to the general public is probably least interesting to scientists.

206 A cubic foot of limestone from some sites can easily yield fifty thousand fossil shells —the remains of small animals living on the continental shelves. This is more of a fossil record than that supplied by all the dinosaurs in all the museums in the world.

207 Some dinosaurs displayed social behavior. Unlike modern reptiles, some dinosaurs seem to have nested in groups and cared for their young. Jack Horner of the Museum of the

This mural, in the dinosaur room at the Field Museum of Natural History in Chicago, was a favorite of mine as a child.

Rockies in Bozeman, Montana, has discovered places where large colonies of dinosaurs nested together, and has argued that they migrated in large herds. This is a new way of looking at dinosaurs, but it is buttressed by Horner's spectacular finds.

208 **Dinosaurs may be related to modern birds.** Some scientists argue that dinosaurs may not be truly extinct because they have living descendants on earth today—the birds. So next time you tie into a turkey dinner, you may be sampling a distant relative of *Tyrannosaurus rex.*

209 **Dinosaurs may have been warm-blooded.** Warm-blooded animals, like humans, maintain a constant body temperature despite the temperature of their environment. Cold-blooded animals, like modern reptiles, do not. It used to be thought that dinosaurs, as reptiles, had to be cold-blooded. Now some scientists argue that they were warm-blooded, like birds.

This argument is made difficult by the fact that soft tissue isn't preserved in fossils. All we have to go on are bones and teeth, so scientists are reduced to trying to make analogies with living animals. It doesn't seem to me that either side of this debate has won any decisive victories yet.

210 **Dinosaurs disappeared suddenly about 65 million years ago.** Perhaps the most interesting thing about dinosaurs is not the way they lived, but the way they died. They disappeared suddenly—in a time that may have been as long as one hundred thousand years and as short as a weekend. At the moment, we can't pinpoint the time much better than that, which is one reason there is such an enormous debate about the extinction of the dinosaurs.

Extinctions

211 **To a first approximation, all species that ever lived are now extinct.** My wife hates it when I say things like that—she calls it "physicists' talk." What I mean is this: the 10–50 million species that are estimated to inhabit the earth today are only about 0.1 percent of all the species that ever lived. The normal course of evolution is for a species to appear, stay around for a while, and then become extinct.

Scientists estimate that over the period represented by the fossil record, species have become extinct at the rate of several hundred per year. You can no more prevent individual species from becoming extinct than you can prevent individual members of the species from dying.

212 **The average lifetime of a species is about a million years.** If we count the advent of humanity with Lucy, the human race is 3 million years old. It seems we're living on borrowed time.

213 **The dinosaurs disappeared in a mass extinction.** When the dinosaurs disappeared 65 million years ago, fully two-thirds of the other species on the earth disappeared with them. For some organisms, such as ocean plankton, the extinction rate got as high as 98 percent. When there is a huge killing like this in the record, it is given a special name—mass ex-

tinction. In light of these facts, the theories touted in supermarket tabloids about the dinosaurs being hunted to extinction by little green men in flying saucers can be safely ignored. Unless, of course, the little green men also hunted plankton!

214 **The mass extinction that got the dinosaurs was neither the most recent nor the worst in history.** Depending on how you count, there are eight to twelve mass extinctions in the fossil record of the past 250 million years. The most recent (somewhat less severe than the one that did in the dinosaurs) occurred about 11 million years ago. The most severe occurred 248 million years ago, at the end of the Permian period. In that one, more than 80 percent of existing species vanished.

215 **The extinction of the dinosaurs was probably caused by a meteorite hitting the earth.** The most recent theory that attempts to explain the extinction of the dinosaurs is that it was caused by the aftereffects of the collision of a meteorite about six miles across with the earth. The dust from the impact would have blocked sunlight worldwide for about three months, killing most plants. In due course, the herbivores would have died, followed by the carnivores. This scenario is called the Alvarez hypothesis, after the father-and-son team of Luis and Walter Alvarez, who first presented evidence for it.

The status of the Alvarez hypothesis is this: there is overwhelming evidence that an impact did, indeed, take place at the same time as the extinction. Whether it caused the extinction or was only part of the cause is still under intense debate.

216 **Mass extinctions may happen regularly, every 26 million years.** New computer data bases of fossils indicate that the mass extinctions are not random, but regular. If this is true, and if one was caused by an impact, then some scientists argue that it's

reasonable to suppose that all of them were caused by impact. This would mean that the earth is bombarded periodically by large objects from space. Why such a bombardment should occur is not at all clear.

217 **Don't worry, the next mass extinction isn't due for a while.** Since the most recent mass extinction was 11 million years ago, we have about 15 million years before the next one comes along.

Human Evolution

218 **The human family tree can be traced by looking at fossils.** Just as the story of the evolution of all living things on the earth can be traced by looking at the fossil record, so too can the history of the human race. The fossil record includes fossils of distant common ancestors of both modern apes and modern humans, as well as more recent ancestors of *Homo sapiens*. The idea that mankind can be thought of as being no different from other animals has always been a hard one for people to accept.

219 **Deciding what a human being is has always been a hidden problem in unraveling human evolution.** This point isn't widely appreciated,

but if you want to trace the human family tree, you have to have a pretty clear notion of what it is that makes a human being different from his or her ancestors. This fact has an important consequence: throughout history, the trend has been to see humanity as special—as different from the rest of nature. Even after the acceptance of Darwinian evolution, an amazing number of mistakes were made because scientists didn't want to see the similarities between *Homo sapiens* and (for example) Neanderthal man (see below).

220 **The most important step in the development of the human race was bipedal locomotion.** What distinguishes humans from other animals is the size of the brain, so it used to be thought that human beings evolved a large brain first, and that our upright walk was something of an afterthought. As it turns out, this is just the reverse of what happened. The walking came first, then the intelligence. The question of why and how hominids began walking upright is still a subject of intense debate among scientists, but it is clear that early hominids had a small brain (perhaps a fourth the size of a modern human's) and walked upright.

221 The belief that mankind acquired intelligence early contributed to the great Piltdown hoax. In 1912 Charles Dawson, an amateur paleontologist, reported that he had found skull and jaw fragments in a gravel pit on Piltdown Common in southern England. The fossil had a large skull (indicating intelligence) and a primitive jaw. After forty years, during which Piltdown man became increasingly difficult to square with developing ideas about human evolution, the fossils were reexamined. The skull fragments turned out to be modern (though cleverly stained) while the jaw was that of an orangutan.

With modern techniques like radioactive dating, primitive hoaxes like Piltdown are unlikely to be repeated. I have to say that when I saw the fossils at the British Museum, with the full value of hindsight, the fact that the teeth had been filed was pretty clear.

Who was the perpetrator? No one knows, but my own favorite candidate is Arthur Conan Doyle, the creator of Sherlock Holmes, and a neighbor of Dawson's.

222 Homo sapiens is the only surviving member of our genus. In the normal course of affairs, you would expect that there would be many other species of the same genus as us—call them *Homo A, Homo B,* and so on. In fact, our family tree has been stringently pruned, and we are the only surviving species, not only in the genus *Homo,* but in the entire hominid family. Whether our ancestors did the pruning by wiping out their competitors, or whether the pruning was done by normal natural selection is an open question.

223 There are large gaps in the human family tree as it is known today. A typical situation is this: we find a few skulls of one species at a few sites in one region, then a few skulls of a different species at sites corresponding to a later geological time in another region. Whether the two skulls represent parallel branches of the family tree, or whether one is the ancestor of the other are questions that cannot be answered from the data alone. Consequently, there is an inordinate amount of argument about the exact set of branchings that led to modern man.

224 You can't tell the players without a scorecard. One of the hardest aspects of studying human evolution is the strange-sounding names attached to various members of the family tree. Here, in order of appearance, are the names of the players, along with explanations of what the names mean:

Ramapithecus ("Rama's ape")—the fossils were discovered in India and named after the Hindu deity Rama.

Proconsul ("before Consul")—in the 1930s, there was a vaudeville act in London featuring a chimpanzee named Consul. In a fit of whimsy, the discoverers called their fossil the precursor of Consul.

Australopithecus ("southern ape") — these fossils were first found in Africa. This is a genus name, and there were several different species of Australopithecus.

Homo habilis ("man the toolmaker")—first fossils to be found associated with stone tools

Homo erectus ("man the erect")

Homo neanderthalensis ("Neanderthal man")—the first "human" fossil to be discovered and recognized was in the valley of the Neander River in Germany

Homo sapiens ("man the wise")— you and I

Cro-Magnon man—same as Homo sapiens—the name comes from the local term for a rock shelter in France where remains were first found.

225 Both modern humans and modern apes descended from a common ancestor. From about twenty to ten million years ago a chimpan-

zeelike animal called Proconsul lived in Africa. This animal is about as close as you can get to having a "missing link," and some scientists claim that it is the most recent common ancestor that we share with the apes. After Proconsul, the family trees of the apes and the hominids (the ancestors of the human beings), began to diverge.

Between fourteen and eight million years ago, an erect walking apelike creature called Ramapithecus lived in Africa and Asia. It shares many features with modern humans, including erect posture and similar jaw structure.

226 Proconsul has had an interesting and checkered career as a fossil. Many of the remains that we know of this particular ape were not recognized as significant when they were first found, and were simply tossed into "miscellaneous" bins in museums. In the end, they were actually discovered in the museums rather than in the field! Makes you wonder about what else is waiting on those dusty shelves, doesn't it?

227 The first great gap in the the human lineage is from eight to three million years ago. We do not know much about what happened to our ancestors after Ramapithecus. This is almost entirely due to the fact that we do not have enough fossils of the early apes and ape-men to sort things out. It is the most poorly known portion of the human family tree.

228 The first true "human" was Australopithecus. If you call members of the hominid family "humans," then members of the genus *Australopithecus* qualify as the first humans. They were erect animals about three feet high, probably fur-covered like modern apes. They were around from about 4 to 1.5 million years ago. There were several species of *Australopithecus*, the oldest being *Australopithecus afarensis* (see the following). Later, two separate species evolved, one robust, strong, and probably vegetarian; the other small, quick, and probably a hunter. All the different Australopithecine species died out at least a million years ago. No one knows why they disappeared, although competition with the direct ancestors of *Homo sapiens* has been mentioned as one possibility.

229 Lucy, the "Oldest Human," was an australopithecine. The earliest and most famous "human" fossil was nicknamed Lucy when it was discovered in 1974. The nickname came from the fact that the discoverers celebrated their find by a night-long campfire party in which the Beatles' song "Lucy in the Sky with Diamonds" figured prominently. Lucy was a young female of the species *Australopithecus afarensis* who lived 3.5 million years ago. (The name means "southern ape from the Afar triangle region of Ethiopia"). We believe that she represents a species that lived in groups and had family units. They certainly walked erect. The discovery of Lucy's skeleton in Ethiopia was perhaps one of the greatest finds in the human fossil record, and it produced one of the most complete skeletons we have of any of our ancestors.

230 The direct line to modern man was through Homo habilis. From 2 to 1.5 million years ago *Homo habilis* lived in Africa. Homo habilis made a variety of stone tools, including tools for cutting and scraping, as well as hammers for forming new tools from flint. They lived in hunting bands and were about the size of a modern twelve-year-old. They also had large brains. Because of their resemblances to modern man, they are classed in the same genus (*Homo*).

231 Most of the famous early fossils were those of Homo erectus. This recent ancestor lived from about 1.5 million to about five hundred thousand years ago. They had larger brains than *Homo habilis* and were not much smaller than modern man. Most important, however, was the fact that they used fire, something that no one before had done.

When people first began discovering the skulls of early men, there were so few of them that each had a special name. They went by names like Java man, Peking man, and so on. As the collections began to grow, and as similarities between the fossils began to be seen, it was realized that all of these different "men" were simply different members of the same species, *Homo erectus.*

232 **There was a time when many "humans" were on the scene.** A million and a half years ago, many members of our family tree may have coexisted in Africa. There could have been two separate genera (*Australopithecus* and *Homo*) and separate species within each. Things must have been interesting. What, for example, happened when a band of *Homo habilis* encountered a band of *Australopithecus?* There should be material for a great novel there!

233 **Neanderthal man was no "Neanderthal"!** It was only a few hundred thousand years ago that anything like modern man appeared on the scene. The first of our kind was the Neanderthal man. Whether or not Neanderthal man was a subspecies of *Homo sapiens* (*Homo sapiens neanderthalensis*) or a separate species (*Homo neanderthalensis*) is a subject of some debate. What is clear is that a hundred thousand years ago Europe and Asia were populated by tribes of beings very much like ourselves.

There is a common misconception about Neanderthal, enshrined in the image of Neanderthal as a shambling, hulking, brutish creature of very low intelligence. In fact, Neanderthal had a larger brain than modern *Homo sapiens.* The shambling walk that we ascribe to the Neanderthal comes from the fact that the first Neanderthal skeleton analyzed was that of a man who suffered from advanced arthritis, and was therefore stooped over. Modern reconstructions of Neanderthal show someone who could probably pass unnoticed in the subways of any major city.

In addition, Neanderthal had an advanced religion, they buried their dead, and toward the end of their reign they made ornaments and other artifacts that we associate with human civilization.

234 **The best explanation of Neanderthal man.** When the first Neanderthal skeleton was discovered in 1856, Franz Meyer of Bonn University declared that the skeleton belonged to a Cossack who had died while pursuing Napoleon across Europe. The man had rickets, said the good professor, which explained his bowed legs, while the pain of the disease caused him to knit his brows, producing a heavy brow ridge. How the rickety Cossack scaled a 100-foot cliff to get to the cave to die wasn't explained.

235 **Neanderthal disappeared suddenly in Europe about thirty-five thousand years ago.** When Cro-Magnon man (us) appeared on the scene in Europe thirty-five thousand years ago, Neanderthal man disappeared. We do not know why this disappearance took place, but we do know that as a result there is only one member of the genus *Homo,* and one hominid on the earth—*Homo sapiens*. Thus, the human family tree can be thought of as a series of experiments, each side branch becoming extinct as a new and more successful model appears on the scene.

Enduring Mystery

236 **What happened to Neanderthal?** There are several theories: (1) Neanderthal was wiped out by vicious invaders (us); (2) Neanderthal interbred with the new arrivals so that the modern human gene pool has a sizable proportion of Neanderthal genes in it; or (3) Neanderthal was unable to compete economically or ecologically with the new arrivals, and therefore became extinct, as so many other species have. At the moment, the last of these seems to be the majority opinion among paleontologists, but fashions in this question change.

237 **What you think about Neanderthal's classification affects how you think he met his end.** If you think Neanderthal is just a subspecies of *Homo sapiens*, then it makes sense to talk about modern man as the outcome of interbreeding between Neanderthal and Cro-Magnon. If Neanderthal is a separate species, it doesn't.

During the 1970s and 80s, the evidence seemed to support the subspecies viewpoint, and you may have learned things that way in school. Lately, however, things seem to be swinging back the other way. The prime new evidence are discoveries of sites in the Middle East where modern man and Neanderthal seem to have lived side by side for tens of thousands of years without interbreeding. If they didn't interbreed because they couldn't, this would be a classic test for different species.

238 **Names in the News.** The people you are most likely to read about in stories on human evolution are members of the Leakey family and Don Johansson. The late Louis Leakey and his wife, Mary, were pioneers in human paleontology, and opened up the famous Olduvai Gorge site in Tanzania. Their son, Richard, is a major figure in his own right. The Leakeys elucidated much of what we know about *Australopithecus* and *Homo habilis*.

Donald Johansson is the discoverer of Lucy, the oldest human fossil. At least as far as the media is concerned, he is engaged in a major battle with Richard Leakey over who has the oldest direct ancestor of modern humans. A quote from one of Johansson's colleagues: "It's hard to trust a paleontologist who wears Gucci loafers."

239 **We may all have the same grandmother.** Scientists who compare DNA sequences for living humans have suggested that all of us can trace our ancestry back to a single woman, christened "Eve," who lived in Africa about 200,000 years ago.

Recent evidence from fossils, as well as criticisms of the original analysis, however, have called this notion into question.

240 Finding the "First Human" is a goal of many paleontologists. Finding the oldest hominid fossils on the

main trunk of the human family tree is considered an important goal by many workers in the field. The reasons have less to do with the scientific value of the discovery than with the visibility, notoriety, and research funding that go along with it. "Pure" science isn't always as pure as some people think it ought to be!

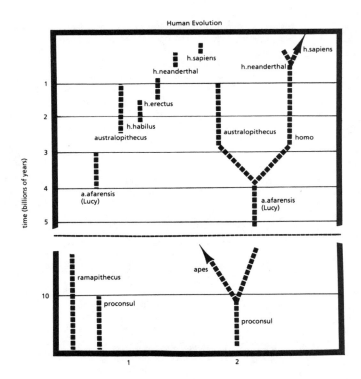

Human Evolution

Here's the family tree—at least, here it is as of the time this book was printed. On the left we summarize the appearance of our fossil ancestors in time, without trying to say who is a branch and who is the trunk.

On the right we show one possible interpretation of the fossil data in the form of a family tree. This is the "Johansson" interpretation, in which Lucy (his fossil) is the ancestor of the entire hominid line. I've also shown Neanderthal man as a separate species.

3

MOLECULAR
BIOLOGY

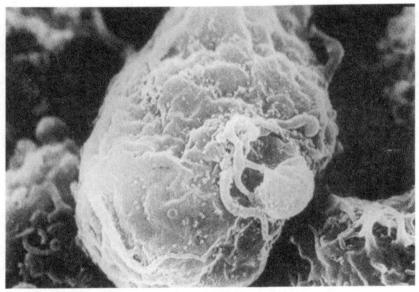

The AIDS virus attacking a human white blood cell.

Molecules of Life

241 **All molecules that appear in living systems are made primarily from just six chemical elements,** namely carbon, hydrogen, nitrogen, oxygen, phosphorus, and sulfur. These are all fairly common elements and would have been available for incorporation when life developed. I am indebted to my colleague Harold Morowitz for pointing out that there's an easy mnenonic for remembering these elements: CHNOPS.

242 **There are four types of molecules in living systems—carbohydrates, proteins, lipids, and nucleic acids.** Each has a different structure and plays a different role. Carbohydrates carry energy and supply some structural tissue. Proteins are the chemical workhorses of the cell (and provide some structural tissue as well). Lipids are important in cell membranes and energy storage. Nucleic acids (DNA and RNA) carry information crucial to the operation of the cell.

243 **Large molecules in living systems are modular and are held together by covalent bonds.** They are built up by combining a specified set of smaller molecules, with large molecules corresponding to different arrangements of the elementary building blocks. Think of a complex building like a skyscraper. It is built up of components—windows, beams, doors, and so on—and different buildings correspond to different arrangements of these components. In the same way, complex proteins are built by stringing together different sequences of amino acids. In this analogy, the role of the mortar is played by covalent bonds, a type of bond formed when atoms share electrons between them.

Carbohydrates

244 **Sugars are the basic building blocks of carbohydrates.** A sugar molecule is built around a ring-shaped structure of carbon, oxygen, and hydrogen. This sketch shows the common sugar glucose—the kind of sugar which the human body uses for energy and which appears in all living cells. There are many other kinds of sugars, all of which have a generally similar structure.

For a fixed proportion of carbon,

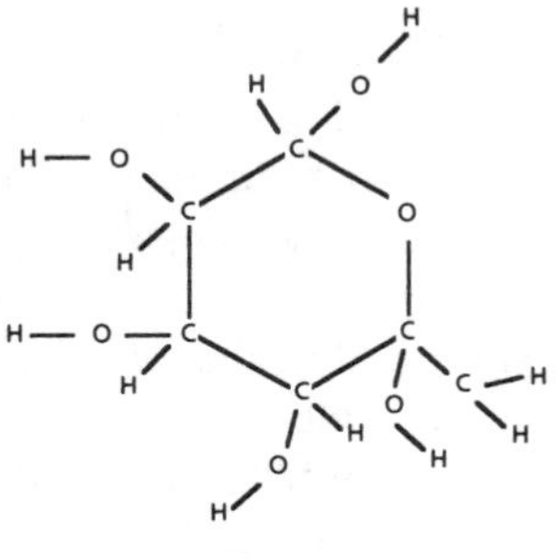

A molecule of glucose.

oxygen, and hydrogen, the arrangement of the atoms can vary as well. (Molecules with the same complement of atoms and different arrangements are called *isomers*.)

Ribose (top) and deoxyribose (bottom).

Another important sugar is called ribose, sketched on top. If one oxygen atom is removed from the ribose, as shown on the bottom, then the molecule is that of ribose without an oxygen, or deoxyribose.

245 Simple sugars combine to form more complex sugars. They lock together when an "H" at one end of one molecule combines with the "OH" at the end of another to form water, leaving behind the two sugar rings linked by a single oxygen atom. Sucrose (ordinary table sugar) is formed in this way by the combination of glucose and fructose (a sugar commonly found in fruits). Chemists call compounds made of two sugars "disaccharides."

246 Starches and cellulose are made from strings of sugars. If we keep stringing glucose molecules together, we wind up with cellulose or starch, depending on which location in the rings we take the "H" and "OH" pairs from. Starch is used as an energy storage compound by living organisms, while cellulose is a rigid molecule that is the principal stiffener in the stems of plants and in wood. Cellulose is also the main fiber of natural cloth—over 90 percent of cotton, for example, is cellulose.

247 Despite their similarities, cellulose and starch have totally different chemical properties. Human beings, for example, can digest starch but not cellulose—that's why we call celery "roughage." Animals like cows must carry around their own bacteria to digest the cellulose for them. The fact that the cotton shirt in your closet, the celery in your salad, and the energy storage system in your body all are made from glucose strung together in different ways illustrates better than anything I can say the wide variety of things that can be made through modular assemblies of small molecules.

248 The term "Carbohydrates" refers to any compound made by stringing sugars together, or any compound that has the composition $C_nH_{2m}O_m$. The term embraces simple sugars (like glucose), compounds made from a few sugars (like sucrose), and those made from many sugars (like starches and cellulose). Chemists use the term "polysaccharides" ("many sugars") for things like cellulose.

Proteins

249 **Amino acids are the basic building blocks of proteins.** These sorts of molecules have a simple general structure: on one end is a nitogen atom with two hydrogens attached (this is called an amine group and gives the compound its name). Then comes a collection of atoms that differ from one species of amino acid to the next, and, finally, at the other end, an "OH" group.

All of the important molecules called proteins are made up of chains of amino acids tied end to end like elephants in a procession. The basic formation of a protein is shown.

When two amino acids come together, the hydrogen from one and the oxygen and hydrogen from another combine to form water. In the process, a longer molecule is left behind which has pieces of each of the two amino acids. An easy way to think of protein formation by amino acids is to think of the amino acids as "squeezing out" the water be-

tween them and then sticking together. This kind of linkage is called a "peptide bond."

The variety of proteins found in nature is made possible by the fact that each different sequence of amino acids corresponds to a different protein. Proteins range in size from less than 100 amino acids (an example: insulin) to hundreds of thousands. The largest protein molecules contain millions of different atoms.

250 **Proteins define our biochemical identity and are the workhorses of the cell's chemistry.** They act as enzymes in all of the complex chemical reactions that go on inside the cells of your body. Sometimes they serve as structural elements—your hair and fingernails, for example, are made from protein molecules.

251 **All proteins in living systems on earth are composed of only twenty different amino acids.** Every protein that appears in any living system on our planet is made of different combinations of the same basic twenty amino acids, despite the fact that many more types of amino acids can be made in the laboratory. The names of the basic twenty are:

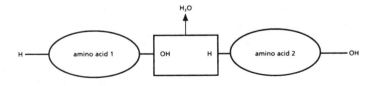

Two amino acids link to start the formation of a protein chain.

glycine	arginine
alanine	asparagine
valine	glutamine
leucine	cysteine
isoleucine	methione
serine	phenylalanine
threonine	tyrosine
aspartic acid	tryptophan
glutamic acid	histidine
lysine	proline

Enduring Mystery

252 **Why these twenty amino acids?** There are two schools of thought. One holds that it is essentially accidental—that the first cell happened to come together with these twenty amino acids in it and then life and therefore all of its ancestors had the same twenty acids. We can call this the "frozen accident" school of thought. The other school of thought holds that there is some law, as yet unknown, which makes these particular twenty amino acids the optimum combination for living systems. We can call this the "biochemical law" school. At the moment, my own leaning is toward the "frozen accident" point of view, but I would not be surprised if its opponents proved correct.

253 **Proteins have a complex, many-layered, structure.** The "primary structure" of a protein is given by the sequence of amino acids along the chain. But a chain of amino acids will not just lie there like a piece of string. Some of the amino acids can form bonds, either with molecules in their own chain or with those in others. As a result of these bonds, proteins acquire a "secondary structure." Some common structures: a corkscrew made of a single molecule (this is what you find in things like hair, fingernails, and wool), separate chains bonded to each other at different points along their length (silk is a good example of this) or separate chains wrapped around each other like cable (as in tendons and cartilage).

In very large proteins, the kind of kinking and twisting associated with secondary structure is established only over some parts of the chain, and you may even have different kinds of secondary structure on different parts of the chain. The entire chain, secondary structures and all, can then fold up into a larger shape, which we call the "tertiary structure." The most important proteins fold into irregularly shaped (but roughly spherical) globules.

254 **Why is silk flexible while wool is stretchable?** In silk the protein chains run in the same direction as the fibers. When you try to stretch the material you are trying to break all the covalent bonds holding the chains together—a tough job. When you fold the material, however, you have to overcome only the weak forces between chains, which requires much less energy.

When you pull on wool, on the other hand, you are stretching out the "corkscrew" in its molecules, an operation analogous to stretching (but not breaking) a spring.

255 The complex, convoluted outer surface of a globular protein molecule makes it ideal for its function as an enzyme. One of the molecules to undergo a reaction will "fit" into one valley on the protein surface, the other molecule will fit in a neighboring valley. The protein holds the two molecules together until the new chemical bonds are formed. The newly formed molecule no longer "fits" on the protein, so it floats off, leaving the protein free to repeat the process. This is how proteins can carry on the chemical business of a cell without themselves being consumed.

Enduring Mystery

256 Why do proteins have the shapes they do? It is a simple (albeit embarrassing) fact that if you tell a chemist the sequence of amino acids in a protein, he or she cannot predict what the tertiary structure of that protein will be. The reason for this failure is not difficult to understand—the protein involves so many interactions between atoms that keeping track of them overwhelms the largest computers. Calculation of protein structure remains one of the major unsolved problems in biophysics.

Lipids

257 Think "fats and oils" and you have a good picture of the third important class of the molecules of life— the lipids. The simplest lipids are built from carbon, hydrogen, and oxygen (though not in the precise ratios found in carbohydrates). Some lipids serve as cell membranes, others as energy storage media, and still others perform a variety of biological functions.

Technically, a lipid is any material that can be easily extracted from living systems and which is insoluble in water. This loose definition explains why so many disparate molecules fall into this classification.

258 Lipids are very efficient at storing energy, dammit! Lipids store about twice as much energy as carbohydrates of equal weight. This is why all animals (and some plants) use them for this purpose. When you overindulge and put on extra weight, chances are that the weight represents food energy that your body is storing as fat, holding it until you need it. Some plants use lipids to store energy (olive oil is an example of a plant lipid), but most use carbohydrates. This is probably because plants do not move, and therefore extra weight does not impose a large burden on their metabolism.

259 Lipids comprise a grab bag of molecules. Cholesterol, testosterone, and estrogen (the male and female sex hormones in humans), vitamin D, and cortisone are all lipids.

Nucleic Acids

260 **The two most important nucleic acids are DNA and RNA, and they are built from nucleotides.** The basic pattern of nucleic acids, like that of proteins, is a repetition of simple building blocks. The building block that is repeated to make up both DNA and RNA is called the nucleotide. As sketched, it is a sugar and a group consisting of a phosphorus atom with four oxygens linked to a kind of molecule called a base. Different nucleic acids use nucleotides with different sugars, and, within a given nucleic acid, successive nucleotides will have different bases.

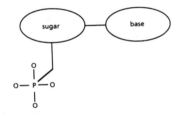

The nucleotide.

Although the individual nucleotide is simple, large molecules can be built from it, just as the biggest skyscraper can be made by repeated addition of various kinds of bricks.

261 **In DNA, the sugar in the basic building block is deoxyribose.** This is what gives the acid its complete name (deoxyribonucleic acid). The base in the DNA nucleotides can be one of four possible choices. These are adenine (A) and thymine (T), guanine (G) and cytosine (C).

The DNA molecule is made from two chains of nucleotides. The bases link to each other across the chain, while the sugars and phosphates link to each other along each chain. The easiest way to think of a DNA molecule is to imagine a ladder. The sugars and phosphates are then the sides of a ladder while the links between the bases form the rungs of the ladder. The "rungs" are, in fact, of only two kinds: a link between A and T or a link between G and C. The structure of the bases does not allow other kinds of links. If you now imagine taking the ladder you've just built and twisting it, you have a molecule of DNA with its famous double helix structure. A typical molecule of DNA will contain millions of nucleotides.

The sequence of bases along the DNA "ladder" is the genetic code.

262 **The RNA molecule is similar to DNA in that it is made from phosphate-sugar-base acid nucleotides.** It

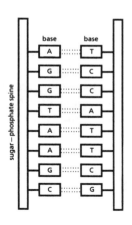

The double-helix structure of DNA.

differs in that it is a single chain (half the "ladder") and the sugar is ribose, rather than deoxyribose. The bases are the same as in DNA except that the thymine (T) which appears in the RNA is replaced by another base called uracil (U) which, like thymine, links to adenine (A).

The Genetic Code

263 There is only one genetic code for all living things on earth. This code, which gives us the fundamental explanation of the laws of genetic inheritance, is carried in the sequence of base pairs in DNA. The bases that constitute the "rungs" in the DNA ladder are what determine the genetic makeup of any organism, with each organism and each individual having a different sequence of those amino acids. Thus, every organism and every individual is different from every other organism and every other individual, even though all have the same kind of molecule at the heart of their reproductive system. You can think of the genetic code as being analogous to other codes in this respect. The Morse code (for example) is a simple sequence of dots and dashes, but an infinite number of different messages can be sent with it. In the same way, the genetic code is simple, but can send the messages that produce both cabbages and kings.

Replication of DNA

264 The first step in transmitting genetic information is duplicating the DNA in the parent's cells. Every living cell has DNA in it, but if that cell is to be reproduced, the amount of DNA has to be doubled so that there is enough for both the parent and the offspring. Here's a step-by-step picture of how it works:

Step 1: Unzipping. A special enzyme moves down the DNA "ladder," breaking the bonds that form the "rungs"—you can imagine this part of the replication process as someone walking along the ladder with a saw cutting the rungs in half. The end result of the unzipping is that there are two separate strands of the original DNA molecule.

Step 2: Rebuilding. Free-floating nucleotides in the cell fluid can hook up to the exposed "rungs" of the separated DNA chains and recreate the missing half of the original ladder. If, for example, there is a free A molecule on the DNA chain, it will naturally attract a nucleotide that has a T molecule attached to it and the two will bond together. The end result of this process is that the particular rung of the original ladder will be replicated.

This process of rebuilding of the rungs goes on on both strands of the original DNA molecule. Gradually, each side reconstructs the pieces of the missing side. The end result is that we have two identical DNA molecules where originally there was only one.

This simple and rather elegant coming apart and rebuilding of the DNA molecule, based as it is on the geometry of four bases, explains everything that we know about genetics. It explains how the reproductive processes of everything from an amoeba to an elephant go on. This example of unity in diversity— unity of process, diversity of form— is one of the most elegant and beautiful truths in science.

265 **The details of the actual process of DNA replication is rather more complicated than the simple process outlined above.** For starters, the DNA does not unzip all at once—instead, enzymes move along the DNA molecule and "unzip" one or more stretches at a time. Those stretches replicate and then the enzymes move on, unzipping as they go. In this way the entire molecule goes through the reformation process without ever having any single molecule come apart completely. In addition, special enzymes operate to form the nucleotides that have been attached to the original strand into a single coherent ladder. The working out of the details of the DNA replication is still one of the major fields of interest in modern molecular biology.

Making Proteins

266 **DNA governs the production of proteins in the cell.** It is the sequence of bases in DNA (the genetic code) that determines which proteins get made, and which therefore determines the way the cell works.

The nature of every protein is determined by the sequence of its amino acids. The "genetic code" is simply the relation between the sequence of the bases in a DNA molecule and the sequence of amino acids of the protein that is created when that part of the DNA molecule acts. In other words, the code translates the information contained in the DNA into the structure of a protein that acts as an enzyme in the cell.

267 **The genetic code is written in triplets.** Extensive experimentation has shown that a string of three bases along the DNA chain determines the position of one amino acid in the final protein. The technical term for a triplet of bases along the DNA molecule is "codon."

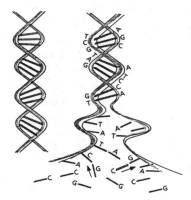

The replication of DNA.

268 Three essential jobs must be done in order to create a protein from a sequence of base pairs in DNA:

1. The information of the DNA must be copied onto some sort of messenger molecule that will carry the information from the DNA to the spot in the cell (usually outside the nucleus) where the actual building of the protein is to take place.

2. There must be some way of transferring the information from the messenger molecule to the protein under construction.

3. There must be some sort of framework that holds all these long and ungainly molecules together while the building process goes on.

All three of these functions are carried out by different kinds of RNA.

269 The first step in the production of the protein is the manufacture of messenger RNA (mRNA). Here's how the first step in the translation of the genetic code works: the two strands of the DNA molecule "unzip" over a certain region along their length. As was the case with DNA replication (see the preceding), the bases that make up the broken "rung" attract nucleotides that are floating around in the cell, except that in this case the nucleotides are those of RNA. The sequence of bases in the DNA are copied as a "negative" onto the RNA, which then floats off. The RNA made in this way is called "messenger" RNA (mRNA) for reasons that will be-

come obvious in a moment, and the process by which it is made is called "transcription."

270 Messenger RNA carries the information from the DNA outside the cell nucleus. Most of the actual work of assembling proteins is carried out in the main body of the cell, not in the nucleus where the DNA is located. The wall of the nucleus in the cell has small holes in it that allow the mRNA molecules to pass through, but are much too small to let the DNA out. The function of mRNA is to allow the cell to carry on the work of building proteins in a different location from the place where the information is stored. You can think of mRNA as being analogous to a computer disk used in a factory. The disk will be prepared in an office by computer programmers, and then will be carried down to the factory floor where it will be inserted into the machines that actually carry out the work, giving them instructions about how they are to proceed.

271 The genetic code is written in the mRNA. Suppose that a particular sequence of bases at one point on a DNA molecule is T T C. The complementary base to T is A, and the complementary base to C is G. Therefore, the sequence in the DNA will show up as the sequence AAG in the mRNA molecule. When the rest of the protein synthesis is carried out, this particular triplet in the DNA will code for the appearance of one particular amino acid (in this case, lysine) in the protein.

Pop Quiz

What will the codon ATG produce in the mRNA? Answer: UAC.

272 The genetic code is redundant. There are sixty-four possible codons that can be constructed from the four amino acids in DNA (4X4X4 = 64). There are only twenty amino acids that go into the proteins of living things. Therefore, the code has to be redundant—some amino acids must result from more than one triplet on the DNA molecule.

Several amino acids are coded for by four triplets in the mRNA, but the champion is leucine, which has no fewer than six triplets devoted to it. For the record, they are UUA, UUG, CUU, CUC, CUG, and CUA.

Pop Quiz

Why should the genetic code be redundant? Answer: For the same reason that spaceships are often built with backup systems: it never hurts to be prepared to deal with a mistake.

273 The assembly of the protein is carried out by transfer RNA. When the messenger RNA arrives at the site where the protein is to be synthesized, another kind of RNA—transfer RNA (tRNA)—comes into play. The tRNA molecule is folded into a key shape as shown. Across the top of the key are three bases of the type found in mRNA. On the tail end of the transfer RNA is a site which attracts a specific amino acid. There are many types of tRNA—one for each of the sixty-four possible codons. Each tRNA molecule is attracted to the appropriate codon along the mRNA. For example, a tRNA molecule with the codon UUC will line up opposite the AAG in the mRNA. At the other end of this tRNA molecule is the site that attracts the amino acid lysine, as shown in the sketch.

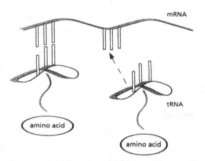

The assembly of a protein. The code written on mRNA (top) is transferred to a sequence of amino acids (bottom) by tRNA molecules.

After a while, each triplet along the mRNA will attract the appropriate tRNA molecule to it, and each tRNA will have the appropriate amino acid at its other end. Through the agency of various enzymes, this string of amino acid is then transformed into a protein. The sequence of amino acids in the protein is completely determined by the sequence of bases in the DNA. Since it is the sequence of amino acids that determines the shape of the protein molecule, and since the shape determines how the protein will behave as an enzyme, the information in the DNA completely determines what chemical reactions occur in the cell, and hence the nature of the cell itself.

The information flow in the genetic code goes from DNA to mRNA to tRNA to protein. The code is simple, although its operation is complex.

274 **Ribosomes hold things together so that the steps of protein synthesis can be carried out properly.** You can think of the ribosome as being like a couple of large spheres stuck together, each with grooves of just the right shape to hold a particular molecule. Every ribosome is custom-made for the production of a particular protein, and

is made of yet another kind of RNA, called ribosomal RNA (rRNA). It serves to lock together around fifty different kinds of proteins into a single massive double-sphere structure on which all protein synthesis takes place.

In a real cell, several different ribosomes may be operating on a protein at the same time—you can think of the protein as being a long piece of tape that is taken up by several spindles at once. You may also have a situation in which one end of the messenger RNA is still being transcribed from the main DNA molecule while the middle part of the mRNA is being read and translated into amino acid sequences at several ribosomes, and the final protein is in the act of folding and forming itself at the other end. In such a situation, all the processes we've talked about in the chain from DNA to protein are going on at once.

Molecular Genetics

275 **Genes are sequences of base pairs on a DNA molecule, with each gene coding for one protein.** When Mendel introduced the gene as the basic unit of inheritance, he had no idea what it could be. Today we identify the gene as a particular segment of a molecule of DNA. A single gene may involve anything from a few dozen to a few thousand base pairs.

The information in one gene is converted into a single sequence of amino acids in one protein. The protein, in turn, acts as an enzyme for a single chemical reaction in the cell. This one gene, one protein rule is one of the central doctrines of modern molecular biology.

276 **There is room for many genes on a typical strand of DNA,** and the exact arrangement of genes varies from species to species. The length of the gene varies according to the complexity of the protein which it codes, of course, and there is often DNA between (and sometimes inside) the genes—DNA whose function we do not yet know. In some organisms, genes even overlap on the DNA chain.

The sum total of all of an organism's genetic code is called its "genome." The human genome contains about one hundred thousand genes. You can get some notion of the complexity of our genetic heritage by noting that each cell in your body contains DNA with enough information to produce one hundred thousand different proteins, each capable of mediating a different chemical reaction.

Because of the difference in the complexity of organisms, not all organisms have the same number of genes. You and other human beings have about one hundred thousand genes. A simple bacteria might have only a few thousand (*E. coli*, for example, has about four thousand).

277 Ninety-five percent of DNA does not code for proteins. Although the basic rule of "one gene, one protein" is the bedrock of modern molecular biology, it is also true that the part of the DNA devoted to genes is only about 5 percent of the total molecule. The rest of the molecule used to be called "junk" DNA, but many biologists now believe that the "junk DNA" isn't junk at all, but contains instructions on when genes

are to be activated, rather than on what proteins the genes make.

278 Each chromosome is a different strand of DNA. Each of the forty-six chromosomes in your cells contains a different DNA molecule—that is, a string of DNA that has a different sequence of base pairs in it. Your entire complement of genes is therefore spread out over all of the chromosomes and is not concentrated on any one.

279 Not all genes are active all the time. When a gene is operating and producing the protein that it is supposed to make, we say that the gene is "expressed." Otherwise, we say that it is "dormant."

At any given time, only a few thousand genes may be operating, with the rest being dormant. For example, every cell contains a gene that would allow it to make insulin, but that gene is active only in cells in the pancreas.

Enduring Mystery

280 How do genes know when to turn on and off during development? Because all the cells in your body arose from the splitting of a single zygote, all of your cells must contain exactly the same genetic information. Yet it is clear that your cells are very different from each other in structure, and they perform very different functions. How can you start with identical DNA and still wind up with very different cells?

We say that the original zygote is "potent" because it is capable of developing into any cell in your body. Later on, the cells become "determined," which means that they are destined to be part of a specific organ and will develop that way no matter what you do to them. Finally, cells become "differentiated" into their present form. The one thing that we do know about this process of differentiation is that is has something to do with the sequences by which the genes are switched on and off during the development of the cell.

In the 1980s, the whole field of gene control was given a shot in the arm by the discovery of a short sequence of DNA in front of a gene that switches on only in the embryo, and is then inactive. Although the sequence (called a "homeobox") was first found in flies, it has since been discovered in humans as well. Scientists can now trace much of the sequence of genes switching on and off as the fly develops, but are still a long way from being able to do so for humans.

Gene Control

281 **The operation of the cell depends crucially on gene control.** The process by which a given gene is turned on and off (i.e., by which a given gene is made to produce or not produce its corresponding protein) is called gene control. The control of genes is important not only in the development and differentiation processes, but in the normal everyday working of the cell as well.

There are many ways that genes can be controlled. The production of a given protein can be controlled by regulating: (1) the rate of transcription of mRNA; (2) the rate at which mRNA breaks down; (3) the rate at which the RNA is converted into protein; and (4) the rate at which the protein molecules break down once they have been created. All of these mechanisms are used in various situations in living cells.

282 **One well understood method of gene control involves the process of controlling the production of mRNA.** In some cases it is possible to find a region of the DNA just in front of the gene which is called the "promoter" region, and which serves as a site onto which a particular enzyme can bind. This enzyme has the effect of preventing the DNA from "unzipping," and therefore inhibits the production of the protein for which that gene codes. When the enzyme is removed the gene will function normally, but when the enzyme is put back in place, the gene is turned off. A gene that has a binding site for a promoter is called an "operon."

The best example of this kind of gene promotion is in *E. coli* (what else?). When it is living in your intestine, the *E. coli* bacterium is in an environment where its source of energy changes drastically over short periods of time. If you drink some milk, for example, the bacterium may occasionally need to have enzymes which allow it to digest lactose, the sugar found in milk. The way the system works is this: the "repressor enzyme" binds to the promoter site in the DNA, preventing the enzymes that would digest the lactose from being made. When lactose starts to appear in the environment, the repressor binds to the lactose and is pulled off the DNA. At this time, the gene becomes "expressed" and enzymes are made to digest the lactose. As these enzymes do their work, the amount of lactose in the cell drops and the repressor eventually binds back onto the DNA and stops the production of the enzyme.

Molecular Biology and Classical Genetics

283 **Modern molecular biology explains the work of Mendel.** Each of Mendel's laws can be related to the workings of individual genes along particular DNA molecules. The acquisition of this knowledge is one of the great achievements of twentieth-century science.

Genetic inheritance is carried in DNA molecules which, in turn, are carried in chromosomes. When an egg and sperm unite, each has half a complement of chromosomes needed to make a normal cell. These chromosomes pair off, and each pair consists of one chromosome from each of the parents. Corresponding genes on each set of DNA molecules

lie opposite each other on the chromosomes themselves, as shown.

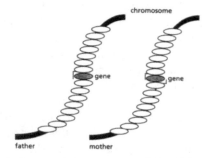

Each gene is a segment of the DNA on a chromosome.

284 **In human beings, the sex of an offspring is determined by the presence of chromosomes known, respectively, as X and Y.** The female ovum (and all the cells in the female body) contain two X chromosomes. The male's sperm are equally divided between X and Y, and all other cells in the male's body contain an X and a Y chromosome. If a sperm containing a Y chromosome fertilizes an egg, the result will be a male offspring, while a female will result from fertilization by an X chromosome. Thus, it is the father who determines the sex of the child, a fact that has not, I suspect, come home to men in many patriarchal cultures, where male offspring are prized and a woman is blamed for bearing female children.

285 **Recessive and dominant traits can be explained in terms of molecular genetics.** One way that recessive and dominant traits work is this: recessive genes do not make a certain enzyme while dominant traits cause that enzyme to be made. Thus if you have one of each kind of gene from your parents, each cell will contain one dominant gene and one recessive gene. In this case, the enzyme will be made (albeit from one chromosome only), so the organism will exhibit the dominant trait. It is only when both chromosomes contain a gene that causes an enzyme not to be made that the recessive trait will appear in the organism.

286 A mutation arises when a strand of DNA is "miscopied" and the wrong base appears in the genetic code. This situation can arise because of the effects of chemicals, radiation, heat, or can simply be spontaneous. Some mutations are not important because of the redundancy of the genetic code. For example, if a particular sequence on the DNA chain is AAT, it will cause the amino acid leucine to appear in a protein. If by mistake this particular codon in the DNA were miscopied to be AAC, it would make no difference to the organism, since this is also a code for leucine.

Manipulating Genes

287 It is possible for scientists to locate the specific spot along a DNA molecule that codes for a specific protein. In other words, we can locate the position of genes along each chromosome. This process is called "mapping." Many of the one hundred thousand genes that make up the human complement of genetic material have been mapped, but the majority have not.

A much more difficult and complex operation involves working out the sequence of base pairs along a DNA molecule. This is called "sequencing." If a particular stretch of DNA has been sequenced, then we know not only *where* the genes are, but, in detail, *what* they are.

288 Biologists have proposed that the entire human genome—all 23 chromosomes—should be sequenced. This would be a decade long, multibillion-dollar project (if it is approved by Congress). It is known as the "Genome Project." The end result would be a complete tabulation of human genetic makeup. It would be the ultimate response to the Socratic dictum "Know thyself."

289 It is possible to "splice" genes into DNA chains by a process similar to splicing moving film. Using special enzymes, a DNA molecule is cut as shown. The fact that the cut is staggered means that another stretch of DNA with the appropriate bases will "stick" to the end of the original sequence. The sticking process is called recombination, and the resulting molecule is called recombinant DNA. In this way, a new gene can be inserted into a string of DNA, and this new gene will be expressed when the manipulated DNA is reintroduced into an organism. Furthermore, the descendants of the first organism to have this gene will also carry it, since the process of cell division begins with the replication of whatever DNA is present.

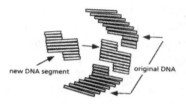

A new gene can be spliced into the DNA on a chromosome as shown. This process is the basis for genetic engineering.

290 "Genetic engineering" refers to the commercial use of organisms that have had their DNA altered by gene splicing. It is a technique of enormous potential. It can, for example, be used to produce bacteria that will secrete useful substances such as insulin and interferon, both of which have important medical uses. It can also be used to produce plants that are resistant to

frost or that manufacture their own insecticides.

Much of the public debate about genetic engineering centers on the danger of releasing hitherto unknown life forms into the environment. In the 1970s, in fact, scientists voluntarily called a moratorium on recombinant DNA research until safety guidelines could be established. These are now in place, and the debate has shifted to specific field testing of the technique, particularly the field testing of genetically engineered plants.

Genetic engineering at work. The large rat had a gene that codes for a growth hormone inserted into its DNA.

The Virus

291 A virus is either the most complicated piece of inorganic matter or the simplest piece of living matter, depending on whether you want to say that it's "alive" or not. A virus consists of a core of either RNA or DNA surrounded by a coating made of protein. Although a virus can sur-

vive independently of a cell, it cannot reproduce without one.

292 Viruses reproduce by using the mechanisms in a cell. When a virus approaches the cell wall, the cell recognizes the protein coating and ingests the virus. Once inside the cell, the protein coating on the virus dissolves, leaving the nucleic acid free to operate in the cell. Some of the normal chemical workings of the cell are diverted into producing material coded for in the viral RNA or DNA. In effect, the viral nucleic acid intercepts the normal process by which the genetic code is transformed into the proteins necessary for cell function. The alien nucleic acid is not only reproduced many times in the cell's machinery, but so is the protein wrapper for the virus. When the cell has reproduced many viruses, it bursts, and the viruses go on to infect other cells.

Enduring Mystery

293 **Where did viruses come from?** Since viruses don't reproduce in the normal way, it's hard to see how they could have gotten started. One theory: they are parasites who, over a long period of time, have lost the ability to reproduce independently. (This isn't so strange—humans have lost the ability to synthesize vitamin C and must now take it in from the environment, even though many other mammals can still synthesize it independently.

294 **Viruses are among the smallest of "living" things.** A typical virus, like the one that causes ordinary influenza, may be no more than a thousand atoms across. This is in comparison with cells which may be hundreds or even thousands of

times that size. Its small size is one reason that it is so easy for a virus to spread from one host to another—it's hard to filter out anything that small. However, even though it is true that, as a general rule, cells are larger than viruses, the range in size of each is so great that there is some overlap—in fact, the largest virus is larger than the smallest cell.

295 **A computer "virus" operates in a manner very similar to a real virus.** Computer "viruses" are small programs that tag along with large programs when they enter a computer and, once inside, take over the mechanism of the computer to perform some function other than the one that's intended. For example, a computer "virus" may simply fill up all the blank memory in the computer with garbage or, in particularly nasty cases, wipe the memory clean. The term "virus" is used because, like real viruses, computer "viruses" do not themselves possess the mechanisms to carry out their nefarious ends. Instead, they commandeer machinery that already exists.

296 **Viruses cannot be killed by antibiotics.** An antibiotic is a chemical that is taken in by a cell and then proceeds to kill that particular cell. It usually works by blocking some vital step in normal cellular chemistry. Because viruses are not cells, this method of attack does not normally work on them. This is why you can't take a pill and get rid of a cold (which is caused by a virus) the way you do to get rid of pneumonia (which is caused by a bacterium).

viral diseases, as are arthropods, but other animals seem not to have this vulnerability.

Enduring Question

297 **Why are viruses so specific?** Viruses seem to be able to invade only one kind of cell, and viral diseases seem to occur in only certain parts of the plant and animal kingdoms. For example, there are many viral diseases of flowering plants but very few of evergreens (gymnosperms). Vertebrates are subjected to a number of

298 **Retroviruses are responsible for AIDS and some human cancer.** This is how a retrovirus works: the viral RNA, aided by an enzyme carried in the virus, produces a strand of DNA that is taken into the cell's nucleus. This DNA then codes for the production of more viral RNA and enzyme, producing new viruses and eventually killing the cell (and the organism).

The Cell

299 **All living things are made from cells.** From the largest sequoia tree to the smallest single-celled bacterium, living things are either single cells or collections of cells. A cell must perform several functions: it must carry out the biochemical business of life, it must generate and process energy, and it must store genetic information to be passed to the next generation. To these ends, cells have a complex structure.

Advanced organisms, both single-celled and multicellular, have nuclei in their cells. A cell with a nucleus is said to be a eukaryote, which translates roughly as "true nucleus."

Primitive cells, on the other hand, have no nuclei, but have their DNA in loose coils in the main body of the cell. A cell without a nucleus is called a prokaryote ("before the nucleus").

300 There are about ten trillion cells in the human body.

301 **Cells come in a wide variety of sizes.** The smallest cells are only a few thousand atomic diameters across. The largest single cell is an ostrich egg, which can be as large as 20 inches long. Most cells, however, are in the range of a few hundred thousand atomic diameters ($10^{-5}-10^{-4}$ meters) across.

There are reasons why most cells

aren't very large or very small. The chemical work of the cell is carried out by proteins, and to make one protein, you need to have a stretch of DNA (i.e., a gene) and two different kinds of RNA. There just wouldn't be room to fit all this machinery into a cell much smaller than the smallest just cited.

For large cells, we begin to run into another problem. As the size of a cell increases, the volume goes up faster than the surface area. Since all of the cell's food must be absorbed and all the wastes excreted through the surface, if the cell gets too big we eventually reach a point where the surface is simply overwhelmed—there isn't enough of it to get the job done. This point seems to be somewhere around the upper limit of the normal cell size.

302 The English physicist Robert Hooke was the first to take one of the new microscopes that his colleagues had developed and looked at the basic structure of living materials. In a piece of cork, he noted a series of hollow, self-contained structures and called them cells. (Today we would call them cell walls.)

The Cell's Energy

303 The universal energy coin is ATP, or adenosine triphosphate. The energy obtained from food or sunlight is used to form this molecule, and the energy stored in it is then made available for other chemical reactions in the cell.

For the record, a sketch of the ATP molecule—arguably second in importance only to DNA—is shown below.

Adenosine triphosphate.

The important thing about this structure is the three phosphorus-plus oxygen groups out in the tail. Each of these is called a "phosphate," and the fact that there are three of them explains the "T" in ATP.

Here's how the cell's energy system works: energy obtained from either fermentation, respiration, or photosynthesis (see the following) is used to tack the last phosphate on to an ADP molecule to make a molecule of ATP. The ATP molecule then moves away, carrying its stored energy with it, until that energy is needed to drive some other chemical reaction. At that point, the extra phosphate is removed (a process that releases energy) and the molecule becomes ADP again. So hooking and unhooking that last phosphate is what keeps the whole living world operating.

Pop Quiz

What is adenosine diphosphate (ADP)? Answer: It's like ATP but it has only two phosphates in the tail.

304 In an average cell there are about two million ATP molecules in use every minute.

305 There are other kinds of short-term energy storage in the cell. In fact, the cell is a little like a modern urban American. We use cash when we're in a grocery store, but credit cards when we want to purchase an airline ticket. In the same way, the cell uses ATP to store the small amounts of energy that are exchanged all the time, but has other processes to deal with large amounts. These processes involve expending energy to pull electrons off some special molecules, then collecting the energy somewhere else when another electron falls back in. The most common cellular "credit card" is a molecule called NAD (the letters stand for nicotinamide adenine dinucleotide and, believe me, you don't want to see a diagram of the molecule).

306 Fermentation is the simplest (and probably the oldest) form of energy generation in the cell. This is a process that involves the breakdown of a carbohydrate molecule (e.g., glucose) into smaller molecules such as lactic acid, ethanol, or carbon dioxide. This breakdown releases energy, which is used to create ATP. It is a relatively inefficient process, producing only two molecules of ATP for each molecule of glucose. Fermentation is carried out in the absence of oxygen; so it is said to be anaerobic. Presumably, cells generated energy by fermentation on the early earth, where there was no oxygen in the atmosphere.

There are many different kinds of fermentation processes, but perhaps the most familiar is the one that leads to the production of ethyl alcohol. In this process, yeast cells convert sugar (glucose) into alcohol and carbon dioxide. *Both* the alcohol and the carbon dioxide are waste products as far as the yeast is concerned—it just wants the ATP. Humans, of course, use the alcohol and let the carbon dioxide go back into the atmosphere.

307 The fact that wine is produced by fermentation has consequences! Fermentation takes place anaerobically, in the absence of oxygen. If wine is left out in the air, fermentation stops and the wine eventually turns to vinegar. If wine has been opened and is allowed to "breathe," the presence of oxygen breaks up some of the delicate molecules in the wine and brings out the full taste and bouquet.

When I was a fledgling oenophile, I remember opening a bottle of Clos de Veugeot and drinking it before the oxygen in the air had done its work. Only with the last glass did I realize what I had done. The thought of that wasted bottle still makes me sad, so I warn you—when you're opening a good wine, *let it breathe!*

308 **Fireflies convert ATP directly to light** In fact, one quick and dirty way biologists have of testing a fluid for ATP concentrations is to drop some ground-up firefly tails into it and see how much light comes out!

309 **In eukaryotes, energy is generated by a more complex process called respiration.** You can think of respiration as "burning" the large molecules by allowing them to combine with oxygen. In this process, a carbohydrate like glucose is broken down all the way to water and carbon dioxide, and all of the energy stored in its chemical bonds is used to made ATP. Respiration is a relatively efficient process, producing up to thirty-six molecules of ATP for each molecule of glucose consumed.

The basic equation of respiration is:

Oxygen + carbohydrate →
carbon dioxide + water + energy

310 **Many cells that normally use respiration retain the ability to run fermentation reactions as a sort of "backup."** When a muscle cell in your body is deprived of oxygen (for example, if it is called upon to work too hard), it has the capability of falling back on fermentation to keep itself going. This "two strings to the bow" strategy is widespread among eukaryotic cells. One exception: the cells of the human nervous system. This is why even a short period of oxygen deprivation can result in serious damage to the brain.

311 **If you work your muscles too hard, the cells will run out of oxygen and start the fermentation process, the end product of which is lactic acid.** The buildup of lactic acid in your muscles results in the all-too-familiar soreness and stiffness the next morning. The solution to the problem is regular exercise, which increases the capacity of the body to supply oxygen to the cells.

312 **Biochemical pathways are the sequence of chemical steps that convert raw fuel to energy in a cell.** In almost all cells, all biochemical pathways lead to ATP, but they can be quite complex and convoluted.

Photosynthesis

313 **Photosynthesis is the inverse of respiration.** The generalized reaction we call photosynthesis can be characterized by the equation

carbon dioxide + energy + water →
carbohydrate + oxygen

The energy used to drive this reaction is, of course, sunlight, and the living things that use photosynthetic reactions are the plants.

314 **Photosynthesis is the basis of all life on**

earth. Plants use the energy in sunlight to synthesize sugars and other carbohydrates. These, in turn, are eaten by other organisms, where the energy stored in them is tapped by fermentation and respiration. All energy in living systems—including the energy you are expending right now to focus on these words—originally comes from the sun through photosynthesis.

315 **Photosynthesis usually involves chlorophyll.** Chlorophylls are a type of molecule that have an atom of the metal magnesium at their heart, a complex ring of carbon and hydrogen around it, and a long tail—they look a little like a kite in the air. The initial step in the process of photosynthesis involves the absorption of a photon by a molecule of chlorophyll. The energy of the photon is used to move an electron in the molecule to a higher orbit, from which it is easy for the electron to move on to another molecule. This "donation" of an electron is the energy input that starts the whole chain of reactions going.

316 **Chlorophyll isn't the only thing in a leaf that absorbs light.** There are two kinds of chlorophyll molecules, absorbing photons corresponding to red and blue light. In addition, leaves can contain other molecules that absorb light and pass the energy on to the chlorophyll. Between the chlorophyll and the pigments, all light except green is absorbed, which is why leaves are that color.

When the chlorophyll stops being produced in the fall, the other pigments determine the light absorption of the leaves, which is why they turn such brilliant colors.

A Dumb Question

317 **Why aren't leaves black?** The most efficient use of sunlight would dictate that all wavelengths of light should be absorbed. Why, then, isn't there a pigment that absorbs green light as there are pigments that absorb yellow and blue-green? If there were, then the leaf would be black. Given the mechanics of natural selection, you would expect such plants to dominate the earth, yet they do not. Is there something in the evolutionary history that precludes the development of a black-leafed plant, or is there a physical reason why such a plant would be inefficient?

318 **Photosynthesis proceeds in two steps.** Once an electron has been removed from a chlorophyll molecule as just described, a complex chain of reactions leads to the production of molecules that store the energy for a short term. These molecules include "ready cash" in the form of ATP and "credit cards" in the form of a cousin of NAD called NADPH. This step in photosynthesis is called the "light reaction."

In the second step, the energy stored in these molecules is used to drive another series of complex reactions that takes carbon dioxide from the air, produces glucose as an end product and oxygen as a waste product. This part of photosynthesis is called the "dark process."

In the absence of light, ATP and NADPH stop being produced and both the dark and light processes in photosynthesis grind to a halt. The name "dark process," then, is something of a misnomer.

319 **Plant cells utilize photosynthetically produced glucose the same way other cells do.** Most plant cells use the process of respiration to process the glucose their chloroplasts produce. As with other types of cells, energy is extracted from the glucose in mitochondria. Thus, plants and animals use essentially the same cellular machinery to get the energy from glucose, and differ primarily in how that glucose is obtained in the first place.

320 **In biology, there is an exception to every rule,** including the rule that photosynthesis requires chlorophyll. In 1971, biologists identified a bacterium that lives in salty environments (a so-called halobacterium). This bacterium does not use chlorophyll but still is capable of photosynthesizing. It produces a type of pigment similar to that found in eye colorings and combines it with a protein to form purple patches in its cell membrane. The patches produce ATP by photosynthesis, and the ATP then drives the metabolism of the cell.

The Structure of the Cell

321 **The cell is not a blob,** but a very complex structure. You can compare a living cell to a refinery or chemical processing plant. Raw materials are being brought in and moved around, thousands of chemical reactions are going on, and the products of those reactions are themselves being carried to other places within the cell or sent back out into the greater organism of which the cell is a part.

Because of this complexity, the term "protoplasm" is seldom used by modern biologists. The term originally meant "living stuff," and referred to what most people believed was a fairly simple fluid inside the cell. Today, the term "cytoplasm" is used to refer to the liquid that exists between the many different structures inside a cell. ("Cyto" is a prefix that means "cell.")

322 **You can think of the cell "factory" as having three major systems.** These are: (1) the set of operating instructions that tells everybody what they're supposed to do; (2) the chemical factories themselves, some of which provide the energy for the cell and some of which fabricate new materials; and (3) the transportation system, which moves materials around from one part of the cell to the other. Part of the transportation system is the membranes that surround individual parts of the cell and separate the cell from the outside. You can think of these as the "shipping docks" that keep out unwanted material and bring in materials the cell needs.

In a typical eukaryotic cell, specialized receptors in the outer membrane shuttle raw materials in and finished products and waste out. Inside, tiny spheres full of chemicals constantly glide along on an intricate three-dimensional system of filaments, carrying their loads to differently shaped objects where the chemical business of the cell is done.

In the nucleus, tangled strands of DNA unzip and send out instructions that are eventually transcribed into proteins.

Cell Membranes

323 **The membranes that separate the cell from its environment and that separate one part of a cell from another are made of a type of lipid (fat) molecule.** These molecules have the property that one end is attracted to and the other end is repelled by water. Left to themselves in a fluid, they will form themselves into a double layer with the side that is attracted to water facing outward and the other side enclosed. Membranes in the cell can be thought of as a stack of these doubled molecules, side by side. The molecules in the stack shift around—in fact, you

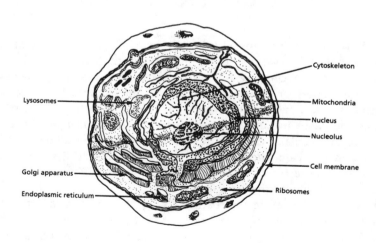

Lysosomes

Golgi apparatus

Endoplasmic reticulum

Cytoskeleton

Mitochondria

Nucleus

Nucleolus

Cell membrane

Ribosomes

Diagram of a cell

can think of the cell membrane as being something like the layer of Styrofoam balls that people sometimes put on their swimming pools to keep them from freezing during the winter. The layer is impervious, yet not rigid.

324 **The cell membrane contains receptors,** which are larger protein molecules imbedded here and there among the lipids. You can think of them as being something like floating basketballs on the surface of a Styrofoam-ball-covered swimming pool. There are many different kinds of receptors, and, like other proteins, they have a complex three-dimensional shape. When a molecule that fits the protein shape comes by, it binds to the protein and is drawn into the cell. In this way, the cell membrane actively picks out the molecules it needs from those in its neighborhood.

325 **AIDS is a deadly disease because the virus responsible for it happens to fit into a receptor that is normally found in the membrane of the human white blood cell.** The receptor, believing it is only doing its job, draws the virus into the cell, with tragic consequences.

326 **Material can be moved across cell membranes in bulk, or one atom at a time.** To move material into a cell in bulk, the membrane indents, folds around the thing to be transported, and then necks off. The result: the material is conveniently encapsulated in a bit of membrane called a vesicle. When the contents of a vesicle need to be moved to the outside (as, for example, when some products of the cell's chemical factories are secreted back into the bloodstream), the vesicle approaches the membrane, opens a small hole, and dumps its contents as if it were squeezing a syringe.

Some small atoms can move across the cell membrane by the simple processes of diffusion or osmosis, but larger molecules often need help. It is the function of the proteins to provide this help. They can do so either by opening a channel through which the large molecule can move, or by actually pushing the molecule through ("pumping"). Both of these processes occur in normal cells.

327 **You know from everyday experience that transport occurs across cell membranes.** You know, for example, that if you put wilted lettuce in water, it "crisps up" by absorbing water into its cells. You know it must be possible to pump materials through cell walls because there are fish (such as salmon) that can live in both salt and fresh water. When a salmon is in fresh water it takes in salt through the cells in its gills. In salt water, on the other hand, it excretes salt through those same cells. In both cases, the salt moves "uphill" and has to be pumped.

328 **On the outside of the cell membrane,**

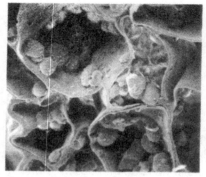

An electron microscope photo of the cell walls in a leaf.

plants, fungi, and some protists have a rigid structure called a cell wall. In plants, the cell wall is made mainly of cellulose and provides the stiffness needed to support the stem and branches. The strength of materials like wood is due to cell walls, which survive long after the cell itself has died.

The Cell's Skeletons

329 **One of the things you'd notice if you went inside the cell is that the entire structure is laced with protein filaments (called a cytoskeleton).** There are different kinds of filaments, and they give the cell its shape and structure. Cells that move do so by shortening and lengthening these filaments or cables, and the cables sometimes protrude outside the plasma membrane to form cilia—little hairs. It is these hairs in cells in the lining of our respiratory systems that constitute our first line of defense against infection.

330 **Your hair and the outer layer of your skin started as cells that had a particularly rich and convoluted set of cables as their cytoskeleton.** When the cells died, the skeleton remained to form these two parts of your body. In a sense, then, you've been touching and manipulating cytoskeleton all of your life.

331 **The cell has a transportation system,** made up mainly from small filaments called microtubules. When a particular batch of chemicals is made up in one of the chemical factories, it is sent out in a vesicle made of the outer membrane of the "factory." These vesicles are then moved along the microtubules, which can be thought of as a system of miniature railway tracks.

With thousands of chemical reactions going on at a time, the job of making sure that the vesicle carrying the right molecules gets to the right location in the cell is not trivial. The "addressing" system is based on the property of specialized molecules on the surfaces of the vesicles. These molecules are recognized by receptors in the wall of the internal "chemical factories." When the right vesicle comes along, the receptor pulls it inside the "chemical factory," just as the receptors in the external cell wall pull in material from outside the cell.

You can think of the cell's transportation system as being like a mail system where letters are sent out at random. You receive all kinds of mail at your house but keep only letters addressed to you, sending the rest on their way. Eventually such a system, inefficient as it may seem, will get all the mail delivered.

Organelles

332 **The organelles are the cell's chemical factories.** An organelle is any organized structure inside the cell. Most of the chemical business in the cell gets carried on in various organelles. If the cell were a large building, the organelles would range in size from beach balls to rooms, and there may be thousands of them in the cell. There are many different types of organelles, each performing a different function.

333 **Mitochondria are the "factories" that supply the cell with energy.** There may be hundreds or even thousands of these sausage-shaped organelles scattered throughout each cell in your body. The carbohydrates, fats, and proteins that you take in in your food and which are partially digested in the intestines are brought into the cell by the receptor proteins in the plasma membrane and ferried to the mitochondria. Inside these organelles, the molecules from the food are broken down (you can think of this process as kind of a slow "burning") and the energy is converted to the molecule ATP, which is the basic coin of cellular energy. In a building-sized cell, the mitochondria would be a bit larger than a living-room sofa.

334 **Each cell in your liver has over one thousand mitochondria churning away.**

335 **In plants, energy conversion takes place in chloroplasts.** As the name suggests, this is where the chlorophyll is located in the cell, and it is the chloroplasts that give the cell (and the leaf) its green color. Their function is to use the energy in light to convert carbon dioxide to glucose, which the cell uses for energy.

In leaves, only chloroplasts are green—the rest of the leaf is colorless. This is why parts of the plant that don't get sunlight (like the roots) aren't green.

Enduring Mystery

336 **Why is a tomato red?** In ripe tomatoes, carrots, and some other fruits, organelles similar to the chloroplasts provide the color. No one knows what function these so-called chromoplasts perform. One theory is that they attract pollinating insects, but it's hard to square that idea with the fact that the orange part of the carrot is underground!

337 **Mitochondria and us: the ultimate happy marriage.** The prevailing theory among biologists is that mitochondria and chloroplasts were once independent cells, leading their own lives. At some time in the past they entered the ancestors of our cells, and the two have been living happily together ever since.

Several pieces of evidence support this view. For one thing, the cell wall around mitochondria and chloroplasts has two layers of lipid molecules, suggesting that these organelles once had their own cell wall. For another, mitochondria and chloroplasts have their own complement of DNA—you can think of them as small prokaryotes living inside the larger eukaryote.

338 By studying the DNA in mitochondria scientists traced the human family tree back to a single woman, "Eve." The idea is that this DNA is not subject to the changes forced by natural selection, and therefore changes only slowly, at a regular rate. Knowing the rate of change in mitochondrial DNA now, and knowing how much the DNA in two individuals has diverged, we can extrapolate back to the time when they had a common ancestor.

339 The endoplasmic reticulum spreads throughout the main body of the cell. If you've ever watched someone deflate a hot air balloon and pack it into the back of a van, you have some idea of what this organelle is like—it is a huge membrane folded over many times on itself.

Part of the endoplasmic reticulum has many ribosomes attached to the outer surface, giving it a rough appearance. It is therefore called the rough endoplasmic reticulum. Proteins made on those ribosomes are for use outside the cell, and the inside of the part of the endoplasmic reticulum that has no ribosomes (the "smooth" ER) provides a space where they can be "finished up" and then stored.

340 The Golgi apparatus looks like a stack of pancakes. It is named after the Italian biologist Camillo Golgi who discovered it in 1898. In a building-sized cell, the stack would be the size of a large car. The number of these organelles in a cell varies in different parts of the body. The function of the apparatus is to do the final synthesis of proteins that the cell is going to excrete. You can think of the Golgi apparatus as a sort of warehouse-light finishing factory in the cell's machinery.

341 The existence of the Golgi apparatus was in question until fairly recently. It turns out that it is very hard to see it in an ordinary microscope, and throughout the first half of this century most biologists put it in the same category as the canals of Mars. It wasn't until the 1950s and the advent of the electron microscope that its existence was firmly established.

342 Lysosomes are the cell's stomach. In human beings, lysosomes contain about fifty different digestive enzymes. Inside the lysosomes partially broken-down food molecules are reduced to simpler chemicals and then sent back into the cell to be converted into energy in the mitochondria. The unused food and enzymes are carried back to the cell membrane and dumped outside the cell. In a building-sized cell, a lysosome would be the size of a chair.

343 **Lysosomes figure in "cellular suicide."** If a cell is deprived of oxygen for an extended period of time, it "commits suicide." The walls of the lysosome break and the digestive enzymes spill out into the cell. In effect, the cell digests itself. In human beings, cell suicide due to oxygen deprivation takes place in the brain after only four or five minutes. This is why even a short period of oxygen deprivation can cause severe neurological problems.

344 **Ribosomes play an important role in the transcription of mRNA into proteins.** There are millions of them in the cell, most of them stuck to the walls of the endoplasmic reticulum. In a building-sized cell, they'd be the size of golf balls.

345 **In most cells (particularly in plants), there are regions that are filled with liquids and appear to have no internal structure.** These are the vacuoles. In plants, they are the places where liquid wastes are stored, and they are responsible for the stiffness in many nonwoody plants. Waste storage is an important function because plants often aren't able to excrete waste to the environment. The stuff just stays in the cell until the plant dies.

The Nucleus

346 **The nucleus contains the genetic blueprint for the cell's operation.** The nucleus is the largest of the eukaryotic cell's organelles. In a building-sized cell, it would be a sphere the size of a room. Like the mitochondria, the nucleus is surrounded by a double membrane. This membrane is punctuated by pores that allow RNA to move outside the nucleus, but keep the DNA inside.

Inside the nucleus are the chromosomes, which are long tangles of DNA and other materials. In humans, there are forty-six chromosomes, with the DNA in each wrapped up in a tangle. These tangled stretches of DNA are what carry the blueprints that run the cell's entire operation.

347 **Human chromosomes are tightly coiled.** Although they fit easily inside the nucleus of a cell that is roughly a thousandth of an inch across, if they were fully stretched out they would be a couple of inches long. Packing the cell with DNA is roughly equivalent to packing a large house full of clothesline. The incredible complexity associated with folding chromosomes into the cell is one argument that people often give that the so-called "junk" DNA must contain instructions for the operation of genes. How else could the cell know where to start unzipping when it wants to transcribe some RNA?

348 Buried at the heart of the nucleus of the cell is an organelle within an organelle called the nucleolus. Its function is to make the ribosomes.

349 Human red blood cells have no nucleus. Alone among all the cells in the human body, red blood cells do not contain a nucleus. They aren't born this way. When they are made in the bone marrow, they have a nucleus like any other cell—how else could they be created? Soon after they are born, however, the nucleus is extruded from the cell. This is why red blood cells are incapable of carrying out repairs and die after a lifetime of about 120 days.

Prokaryotes

350 Unlike its complex eukaryotic cousin, the prokaryotic cell is a relatively simple thing. It has three parts—a cell membrane, a few thousand ribosomes, and a relatively clear area of the cell where the DNA is located. All of the organelles and complex machinery of the eukaryotic cell are absent.

351 Prokaryotes probably evolved first, with the more complicated eukaryotes coming later. As far as we can tell from the (very sparse) fossil record, there were prokaryotic forms of life some 3.6 billion years ago, while eukaryotes have been around for only a little more than the past billion years.

Enduring Mystery

352 Where are the missing links? The differences between prokaryotic and eukaryotic cells are striking, to say the least. But if the latter evolved from the former, why are there no intermediate stages between the two? Why, for example, are there no cells with loose DNA and organelles? If the evolutionary line really went from prokaryotes to eukaryotes, and we have many living samples of each, why did none of the intermediate stages survive?

Dumb Question

353 Couldn't it be done more simply? When you look at the complex structure of a cell, Rube Goldberg springs to mind. This leads to the question just posed. Is the complexity of the cell due to the long evolutionary history, or is this really the most efficient structure capable of doing what a cell does? As far as I know, biologists haven't even begun to address this question.

The Division of Cells

354 **Right now cells in your body are dividing at the rate of millions per second.** They don't all divide at the same rate, however, and some divide only during growth and then stop. The fastest dividing cells are those in the lining of the small intestine—they divide every few days. In contrast, the cells of your nervous system, once they reach maturity, don't divide at all. In between these two extremes we have cells that divide every few weeks, like those in the skin.

Enduring Mystery

355 **What makes a cell divide?** Why should cells divide rapidly in a child, then slow down or stop in adulthood? This is a question of more than academic interest, because many scientists feel that cancer results from a failure of the mechanism that tells cells when to stop dividing. As evidence, they point to tumor cells that will double in number every few days when grown in culture, but that produce tumors that double in size only over periods of months or years. Even in a tumor, it seems, there is a mechanism that restrains the propensity of cells to multiply.

356 **Chromosomes are where the genetic material of the cell is stored.** In the early days of microscopes, biologists watching cell division noticed that just as a cell began to divide, short stringy objects suddenly appeared in the nucleus. These objects would absorb a colored dye that would make them more visible, and hence they were called "chromosomes."

Today we know that chromosomes are where the cell's DNA is stored, and we recognize that the duplication of the chromosomes is an essential feature of cell division.

But chromosomes contain more than DNA. Although the details are not completely settled, the chromosome in a eukaryotic cell is a fairly complex structure. The strands of the DNA double helix appear to be wrapped around a series of "spools" made of protein molecules, with each strand of DNA wrapped around many different spools.

357 **If the DNA molecules in the single chromosome in E. coli were blown up to be the thickness of ordinary clothesline, it would be five miles long.** This illustrates the point made earlier about the necessity of having instructions coded into the DNA so that the genes can be found.

358 **During the start of cell division, the chromosomes coil up and thicken.**

Most of the time, chromosomes and their DNA are strung loosely around the nucleus. It is only when cell division is about to start that they coil up and become visible through a microscope. This fact led to an interesting question for nineteenth-century biologists—where did the chromosomes go when they weren't visible between cell divisions? With the advent of better microscopes, this question can now be answered—they're there all the time.

359 **Different species have different numbers of chromosomes.** Human beings have forty-six (i.e., twenty-three pairs), mosquitoes have six, dogs have seventy-eight, goldfish have ninety-four, and cabbages have eighteen. There seems to be very little correlation between the complexity of an organism and the number of chromosomes it has.

360 **The organism that has the most chromosomes is a species of fern,** *Ophioglossum reticulatum* that has no fewer than 1260 (630 pairs). The least number of chromosomes in normal cells is found in a species of Australian ant, *Myrmecia pilosula*, whose workers' bodies are made of cells with only one chromosome apiece.

361 **The process of one cell splitting into two daughters, each identical to the original, is called mitosis.** This is the "normal" method of cell division, used by all the cells in your body except for those involved in producing sperms and eggs.

The first step in mitosis is the copying of the DNA. As duplicate sets of chromosomes are made, they pair up so that the two identical chromosomes are linked together in what usually looks like an X-shaped assembly.

While the DNA is being copied, the cell retains its normal appearance. The real action begins when the chromosomes coil up and become visible and the membrane surrounding the nucleus disappears, exposing the nuclear material and making it part of the cell at large.

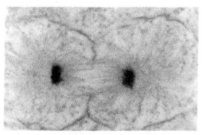

A cell caught in the act of dividing. Notice the formation of the spindles.

362 **The final stage of cell division is the actual fission.** You can think of the chromosome pairs that are linked together after duplication as being like pairs of socks in a drawer, with each pair tied together. There is a cell mechanism (poorly understood at the present) which separates these pairs. In effect, this mechanism reaches into the "drawer," pulls out a pair of chromosomes, and then separates it. While this is going on, a structure called a spindle forms in the cell. The spindle is made of proteins, and looks like lines of "longitude" drawn on the cell, each line starting at one pole and ending at the other. Each member of the chromosome pair is pulled in opposite directions along the spindle so that the two matched chromosomes wind up at opposite ends of the cell. When this process has been completed for all chromosome pairs, there are sets of identical chromosomes at opposite sides of the cell.

Once the chromosomes are separated, another protein framework forms circumferentially around the cell and begins to squeeze in. This causes the cell to divide. The process is similar to taking an inflated balloon, wrapping your fingers around the middle, and then squeezing.

363 During cell division, the operation of genes shuts down. While the duplication process is going on, the normal operation of the genes is suspended. This means that the cell has to give out all the instructions for completing the division (including producing and assembling the proteins for the various structures involved) before the division actually happens. In human cells division typically takes about seven hours.

364 The life history of a cell is frequently represented by a circle like the one shown. The crucial events in cell division occupy different regions of the circle, and the cell actually cycles through all of them, like a second hand sweeping around a clock face.

The start of the division process is signaled by the beginning of the replication, or synthesis, of DNA. This is represented by the "S phase," as shown. Then there is a gap in time (called "G2") before the cell actually undergoes mitosis (M). After mitosis, there is another gap during which the cell carries out its normal functions but produces no new DNA. This is "G1."

Every cell cycles around this "clock" throughout its lifetime, with each "tick" corresponding to the production of one daughter.

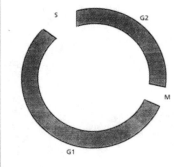

The cell cycle.

365 The divisions that lead to sex cells are somewhat more complicated than normal mitosis. In the formation of a sex cell, we do not want to have an exact reproduction of the original cell. Instead, we want to have a cell with exactly half the normal number of chromosomes. The kind of division that begins with a normal cell and winds up with a cell with half the normal number of chromosomes is called *meiosis*. The beginning of meiosis is similar to the beginning of mitosis—that is, the chromosomes duplicate themselves and form into pairs. Once again, a spindle framework forms, but this time chromosome pairs (rather than individual chromosomes) are pulled off to opposite sides of the cell. Once this process has been completed, another set of spindles forms at right angles to the first and the chromosome pairs are separated. The cell then divides four ways so that the end result is four cells, each with half the number of chromosomes of the original cell. The new cells are the sperm or eggs—the sex cells.

366 Prokaryotes reproduce by fission.

As befits their status as the primitive precursors of the more complex eukaryotes, the chromosomes of prokaryotes are simply loose coils of DNA, without any protein.

The DNA in a prokaryote is usually attached to the outer membrane of the cell. When the DNA is replicated, the new chromosome also attaches to a point on the membrane. Cell division is then accomplished by splitting the membrane off between the attachment points.

4

CLASSICAL PHYSICAL SCIENCE

Isaac Newton, using a prism, discovers the nature of white light.

Classical Optics

367 **Light is a wave.** There are two fundamental ways of transferring energy, one of which is through the use of a wave. Light is a wave, as we can tell from the fact that light can exhibit interference. Unlike other waves, however, light can travel in a vacuum.

368 **Light is a particle.** In this century we have understood that light sometimes also shows the properties of a particle. The first evidence for this point of view came from Albert Einstein's explanation of the photoelectric effect. The particle that corresponds to light is called the photon.

369 **The color of light depends on its wavelength,** with the longest wavelength corresponding to red, the shortest to blue and violet. This hierarchy of colors also corresponds to a hierarchy of energies. Blue light is the most energetic kind of light, while red light is the least energetic.

370 **White light is made from a mixture of all the colors.** When Isaac Newton set out to determine the nature of light, he allowed sunlight to fall on a glass prism that bent each wavelength through a different angle, creating a rainbow of colors. Once he had broken the sunlight down into its constituent colors, Newton put another prism into the apparatus and "reassembled" the beam, producing white light.

371 **Things are colored because of the way that light interacts with the atoms.** When we say that we "see" something, what we mean is that light has traveled from that thing to our eye, where it has triggered a complex chemical reaction in the retina. Thus, what we perceive as color arises from the interactions of light with atoms in the thing we're seeing.

The colors you actually perceive depend on more than simply the wavelength of light. Color vision depends on what other colors are in the visual field and it may even depend on your state of mind. When we see a color, we are seeing the result of a very complicated amalgam of the incoming light, the physiology and nerve connections in our eye, and the processing of nerve signals by the brain.

372 **When objects emit light, electrons in atoms are making quantum jumps.** This means that specific atoms or molecules in the material are sending out radiation with well defined energies and wavelengths, and we are perceiving this radiation as color. Obviously, the energy needed to produce this radiation must come from somewhere. One common way for an atom to acquire the energy to emit radiation is for it to be heated. This is why flames often appear to be colored.

373 You can see atoms starting to emit light when you watch a campfire —just look closely and notice the small colorless gap between the wood and the flames. In this gap, gases are rising and being heated, but have not yet reached the temperature where they can combine with oxygen. It is only after this happens that the atoms acquire enough energy to emit light and produce a flame.

374 For light to be reflected, it must first be absorbed, then reemitted. The light that comes out may be at the same wavelength at which it was absorbed, or at a different wavelength. When we look at the object, then, we see light that is characteristic of the atoms in the material, and not necessarily of the light that is striking it. This is why grass and brick appear to be different colors even though they are both illuminated by sunlight.

375 The colors of absorbed light determine the color of many materials. When light is absorbed by an atom, many things can happen to the energy. One common outcome is that the energy is simply absorbed by the material—it can, for example, be converted into the kinetic energy of the atom, which we interpret as heat.

The colors that aren't absorbed are what we perceive. Chlorophyll, for example, absorbs photons in the red and blue parts of the spectrum and converts their energy into food for the plant. It doesn't absorb green, which is why leaves appear to be that color.

376 Fluorescent materials give off bright colors only when irradiated by ultraviolet light. What happens in a fluorescent material is that ultraviolet radiation is absorbed by atoms, but the atoms then reemit the energy as visible light. The radiation may be reemitted immediately, in which case the materials fluoresce only as long as the ultraviolet source is present. This is what happens with the "black light" so popular in nightclubs. It may also be emitted hours and even days later, in which case the material will "glow in the dark." The fact that in fluorescence we are looking at light emitted directly by atoms, rather than what is left after light has been subtracted, explains why these colors appear so vivid and bright to us.

377 When light is transmitted through a material, only some wavelengths are absorbed. Depending on how atoms are arranged in a material, the transmitted colors may or may not be the same colors that are reflected by the thin layer of atoms at the surface. This explains why some materials (e.g., a leaf) look different from underneath than they do when seen in reflected light.

Pop Quiz

Why do stained-glass windows look gray from the inside of a church at night but bright from outside?: Answer: From the inside of the church you see reflected light, from the outside you see transmitted light.

378 The sky looks blue because of the way light scatters from molecules in the air. When light encounters atoms and molecules, blue light is scattered more readily than red. This means that when white light from the sun comes through the atmosphere, the blue component of the light is scattered out of the beam, but the red component is not. This is why the sun looks yellow (yellow = white minus blue) and why the sky is blue (it is light that has been scattered to our eye from molecules in the air).

Particulate matter in the air (such as smoke or dust), scatters all wavelengths equally, so that sunlight scattered from such material looks white. The presence of these particles accounts for haziness in the sky and the pale ring around the sun on bright days.

379 A polar bear's fur appears to be white because it has many tiny air bubbles in it for insulation. These air bubbles, like dust particles in the air, scatter incoming light and make the fur appear white. The fur fibers themselves are colorless. In the same way, a newborn baby's eyes are blue because small particles of material in the iris scatter blue light preferentially, like molecules in the air. A baby's eye color may change after several months as his or her body starts manufacturing the pigment that will eventually color the eyes.

380 Sunlight and light from a fluorescent bulb may look the same to the unaided eye, but in fact they contain slightly different mixtures of the different wavelengths that make up white light. Thus, the light that is reemitted by objects illuminated by sunlight and fluorescent bulbs is slightly different. This is why clothes sometimes appear to be different colors in the store than on the street.

Optical Instruments

381 The purpose of a lens is to bend and focus light waves. In a typical lens light hits the front end of a curved piece of glass, is refracted, and then is refracted again as it leaves the back end. The lens forms an image which is, in general, a different size from the object being examined, so that the object may appear larger or smaller than it really is.

Lenses can be convex (i.e., the glass can bulge out) or concave (i.e., with the glass at the center thinner than at the edge). If light rays from a distant source enter a convex lens, they are brought together at a single point called the focus, and the distance from the lens to the point at which light is brought to focus is called the focal length. The fatter the glass, the shorter the focal length and the more powerful the lens will be.

382 **The lens in the human eye, unlike a lens made of glass, can be made to change its focal length.** A ring of muscles around the lens can contract and thicken the lens. The closer something is, the more the muscles contract, and the thicker the lens becomes. When you look at something farther away, the muscles in the lens in your eye relax and the lens gets thinner.

383 **Refraction depends on the wavelength of light,** so a piece of glass will bend red light through a smaller angle than green light, and green light will be bent through a smaller angle than blue. This means that the images formed by the lens (as well as the focal point) will be at slightly different places for different colors. Thus, the blue part of an image might be brought into sharp focus while the red part is still fuzzy. Unless corrections are made, this can lead to a situation (known as chromatic aberration) in which the image formed by the lens seems to be fuzzy and surrounded by rings of a different color.

384 **Rainbows are formed by the refraction of sunlight through falling raindrops.** When you stand with your back to the sun, light enters the front of raindrops, bounces off the back, comes back out the front and into your eye. In the process, the beam is split into its component colors by refraction, and different wavelengths are concentrated at different angles. The different colors you see in the rainbow actually come from different raindrops, with the raindrops that send blue light being closer to the ground than those sending red.

385 **In the 1700s, when astronomers first began building large telescopes, they used mirrors rather than lenses to form images.** The reason for this was that no one knew how to correct for chromatic aberration. Consequently, those who wanted a sharp image of the stars were forced to use reflecting telescopes (see the following) with mirrors made of metal. In the nineteenth century, they learned to correct for chromatic aberration and built telescopes with big lenses, and only in this century have they gone back to reflectors.

386 **A typical simple microscope consists of two lenses that together form a virtual magnified image of a small object.** Historically, most of the discoveries of single-celled organisms in biology were made with just this kind of microscope.

The size of the smallest object you can see through a microscope depends to some extent on the quality of the lenses, but there is a fundamental limit: you can't see details smaller than the wavelength of the light you're using. The smaller the object you want to see (or the finer the detail on something large), the smaller the wavelength required.

387 **You can make a microscope using electrons instead of light,** since electrons can behave as waves. The difference is that electron waves have very short wavelengths, so electron microscopes can provide much finer detail than can light microscopes. In electron microscopes, magnets take the place of glass for moving the electron beam around, and pictures are built up in a way similar to the techniques used in television.

388 **Scanning Probe Microscopes (SPM) can produce "pictures" of individual atoms.** The most developed SPM, called the Scanning Tunneling Microscope (STM), works like this: a sharp stylus is held near the surface of a material and an electric current flows into the electron clouds of the atoms that comprise the surface. The smaller the gap between the stylus and the surface, the higher the current. By making repeated passes over the surface, a picture of the location of the electron clouds

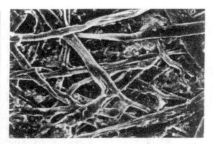

Ordinary things can look strange when seen through an electron microscope. These are fibers of cellulose in a sheet of paper.

(and hence individual atoms) can be built up.

389 **The prime optical instrument of astronomy is the telescope.** Throughout the nineteenth century the most common telescope was of the type shown on the left—light was collected in a large lens and run through a series of eyepieces for observation. This is called a refracting telescope.

Since the turn of the century, telescopes have been of the type known as reflectors (shown on the right). In such telescopes, large concave mirrors collect light and focus it on a mirror that brings the light rays to an eyepiece, which forms the final image. The shift to reflectors was caused by the inability of optical en-

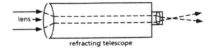
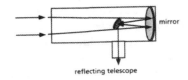

Refracting and reflecting telescopes.

gineers to make high-quality lenses big enough to collect light from very faint distant galaxies. The telescope on Mount Palomar in California is the largest productive reflecting telescope built in this way.

390 **High-tech telescopes are the wave of the future.** The next step will be to marry modern high-speed electronics to optics to build a new generation of telescopes. These telescopes will be a collection of small mirrors, each electronically controlled to keep images in focus. By combining images from each of the smaller telescopes electronically, we can produce a composite image whose quality is equivalent to that which would be produced by a very large single mirror.

This technique was pioneered by the Smithsonian multiple mirror telescope in Arizona in the 1970s, and will be used, in one form or another, in most new telescopes. The Keck telescope, recently built by Cal Tech in Hawaii, has a collection of small mirrors that look like a layer of potato chips, but the mirror is, in effect, twice as big as Palomar.

Waves

391 **Waves are one of two fundamental ways of transferring energy.** If you want to transfer energy from one point to another, there are only two ways you can do it. Suppose, for example, that you want to stand at one end of a room and knock over a milk carton at the other. You can throw something (e.g., a baseball) at the carton and knock it over. This is an example of transferring energy by means of a particle.

Alternatively, you could line up a bunch of bowling pins one next to the other between you and the milk carton and then knock over the pin nearest you. This pin would knock over its neighbor, which would then knock over its neighbor, and so on, until the disturbance reached the carton. In this case, the energy is carried to the carton not by any single particle but by the wave of falling pins.

392 **The motion of a wave is not the same as the motion of the material on which the wave moves.** When you're floating in the surf and a wave goes by, you move up and down, and the water in your neighborhood moves up and down with you. The wave, on the other hand, moves toward the beach. Thus, the motion of the wave (toward the beach) and the motion of the water (up and down) are not the same.

If, as is the case in water, the medium moves perpendicular to the direction of the wave, we say the wave is transverse. If the medium moves in the same direction as the wave (as it does for sound and some seismic waves), we say the wave is longitudinal.

393 A wave is characterized by frequency, wavelength, and speed. The wavelength is the distance between crests, the frequency is the number of crests that go by per second, and the speed of the wave is the velocity of a single crest.

The standard unit of frequency is the Hertz (Hz). For a wave with a frequency of 1 Hz, one crest passes a fixed point each second. The unit is named after Heinrich Hertz, the German physicist who discovered radio waves.

Interference of Waves

394 The property of waves that distinguishes them from anything else is interference, which can be seen whenever waves from two different sources come together. For example, two motor boats might be racing on a smooth lake, each leaving a wake behind it. There will be places in the lake where the wake of the two boats overlap, and this is where you will see interference.

What happens when two waves come together depends on the relative positioning of those waves. I like to think of it this way: imagine that each wave is giving instruction to the surface of the water. If two crests arrive at the same time, then each crest gives the instruction "move up two inches." The result: the water moves up four inches—twice as far as it would for each individual wave. This is called constructive interference.

If, on the other hand, the two waves arrive so that the crest of one coincides with the trough of the other, then the water is getting the instructions, "move up two inches" from one wave and "move down two inches" from the other wave. The result: the water doesn't move at all. This is an example of destructive interference.

If the waves are arriving with a relationship intermediate between these two extremes, then the result is that the water moves up less than it would for constructive interference but more than it would for destructive.

395 The existence of interference distinguishes waves from particles. Two waves can come together at a point and produce the total absence of waves. Two particles cannot do this—you could never, for example, throw two baseballs together and have no baseballs as a result.

396 If the two waves coming together are light-waves, then destructive interference results in dark spots. You may see these when you look at a streetlight through a screen window and see a "cross" with alternating light and dark spots in the arms.

"Dead spots" in an auditorium are places where sound waves (for example, from the stage and reflected off a wall) interfere destructively. Once you're at a concert, there's no cure for this particular problem except to move to a different seat.

397 **The wave nature of light is established by experiments that show that light exhibits interference.** In one famous experiment, British physicist Thomas Young (1733–1829) shone light on a screen with two slits in it (as shown). Waves from the slits move toward the screen and interfere when they come together.

The results of the experiment are shown on the right of the sketch. Midway between the two slits on the screen, the waves come together, interfere constructively, and produce a bright line. Moving away from this central bright line in either direction we see a series of dark and bright lines, corresponding to successive regions of constructive and destructive interference. You could not get a bright central line result if there were particles (like baseballs) coming in.

398 **Waves can bend around corners.** If you've ever flown into a city with a harbor on a sunny day, you've probably seen something like what's sketched. Waves from the lake or ocean come into the breakwater, and then from the breakwater move into the harbor in all directions. This means that someone standing off to the side of the breakwater will see a wave, even though there's no way that the wave can get to his or her position by traveling in a straight line. This ability of waves to go around corners is an example of the phenomenon known as diffraction, and constitutes yet another difference between particles and waves.

399 **When waves encounter a surface, they can be reflected, refracted, or absorbed.** If it is refracted, the wave changes direction. The easiest way to think about this is to note that in glass light moves more slowly than it does in air. A wave front approaching the glass, then, is like a line of soldiers marching along and encountering a marsh which slows them down. As each soldier enters the marsh, he slows down and the

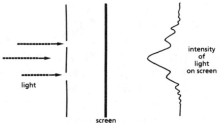

light

screen

intensity of light on screen

A typical experiment showing the interference of light waves.

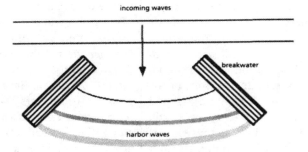

incoming waves

breakwater

harbor waves

Waves bending around corners.

net effect is that the entire line wheels around into a new direction. In just the same way, light changes direction when it enters glass or some other material.

Waves can be refracted without actually encountering a sharp surface. For example, seismic waves moving through the earth encounter rocks of gradually changing density and composition. They, too, change direction because of refraction but do so continuously, rather than all at once. Consequently, they follow a curved path through the earth.

effect" after the Austrian scientist Johann Christian Doppler (1803–1853). If the source of a wave is moving, a spherical wave emitted by the source at any moment moves outward in a sphere which is centered on the spot where the emitter was when that wave was emitted. Thus, the sphere we have labeled "A" in the sketch was emitted when the source was at A, the sphere we have labeled "B" was emitted at source B, and so on. The net effect of the motion of the source, then, is that someone standing in front of the source (i.e., someone watching

400 When you see something that looks like a puddle of water on a hot highway, what you are actually seeing is light that was moving from the sky toward the highway, but has been bent around by refraction in the air until it comes to your eye.

401 One of the most important wave phenomena is called the "Doppler

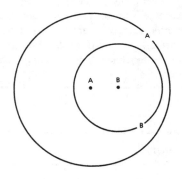

The origin of the Doppler effect.

the source move toward him or her) will see the crests bunched up, while someone standing in back of the source (i.e., who sees the source moving away) will see them stretched out. If the wave is sound, the person standing in front of the wave will hear a higher pitch, and the person standing in back will hear a lower pitch. If the wave is light, the person standing in front will see the light shifted toward the blue while the one standing in back will see it shifted toward the red.

Pop Quiz

Give an example of the Doppler effect that you have experienced. Answer: The sound of a car horn or engine changing pitch as it moves by you on the highway.

402 Police radar units send out radio waves that are absorbed by the atoms in your car. The energy of the waves is then reemitted by the atoms and detected by the radar apparatus. The wave that your car emits will be Doppler-shifted because your car is moving. Thus, the contributions of Johann Doppler to science are now used for, among other things, regulating traffic on congested freeways.

Award

The Casablanca award for the classiest experiment in history. The movie *Ca-sablanca* is often called the classiest movie ever made. In this spirit, we offer the Casablanca award to the first experimental verification of the Doppler effect. This was done by the Dutch scientist Christian Buys Ballot (1817–1890). He assembled a group of trumpeters and put them on an open railroad car. Standing along the railroad tracks were a group of people who had perfect pitch (i.e., who could tell what a musical note is simply by listening to it). The engine was started up, the train went by, the trumpeters played their notes, and the observers recorded what they heard, verifying Doppler's predictions in the process. Next to this, experiments with oscilloscopes and microchips seem pale and insipid.

403 Sound is a wave. When your vocal chords vibrate, they alternately push air molecules closer together and pull them farther apart, and this motion starts the wave.

When a sound wave passes, the air molecules move back and forth in the direction of the wave, rather than up and down as in water waves. In regions where the molecules are packed more densely, the pressure is higher. This is the crest of the sound wave. When the wave reaches your ear, it pushes on the eardrum and initiates the process we call hearing.

The higher the frequency of the wave, the higher the pitch we perceive. The higher the pressure of the crest, the louder the sound we hear.

Electricity and Magnetism

404 **There is a connection between static cling in your clothes and a magnet holding a note to your refrigerator door,** as there is between all electrical and magnetic phenomena. In fact, the discovery of this connection marks one of the highlights of nineteenth-century physics. What we now believe is that electricity and magnetism are simply different aspects of the same fundamental force, which we call the electromagnetic force.

405 **Electrical current can produce magnetic effects.** If an electrical current runs through a wire, a compass near that wire will be deflected. In other words, an electrical current—moving electrical charges—can produce a magnetic field. This is the very first connection that was discovered between electricity and magnetism.

Actually, it was discovered almost in spite of the experimenter. Hans Christian Oersted (1777–1851) was a Danish physicist who loved to give lecture demonstrations for his students. One day while he was messing around at a big desk full of equipment in front of a class, he noticed that whenever he connected a battery to a circuit, a compass needle in the vicinity moved. This accident was the basis for what may have been one of the most important practical discoveries ever made.

406 **The great advantage of an electromagnet is that it can be turned on and off.** An electromagnet is a loop (or loops) of wire, through which electrical current flows. The current produces a magnetic field—the more loops in the wire (and the stronger the current), the stronger the field. Thus, the loop of wire behaves just like an ordinary magnet, and can pick up pieces of metal with the best of them.

The strength of the magnet can be adjusted by regulating the amount of current flowing in the wire. For example, in junkyards you often see magnets used to lift cars. When the current runs through the magnet, the car is attracted and the magnet, usually suspended from a crane, can lift the car up. When the operator wants to drop the car, he turns off the current. As soon as the current stops, the loops of wire stop acting as a magnet and the car stops being attracted to them. The force of gravity (which was always there) now takes over and the car falls.

407 **Magnetic fields can cause electrical effects.** This is another connection between electricity and magnetism. If you move a magnet around near a loop of wire, or if you rotate the loop near a magnet, an electrical current will flow in the loop, even though there is no voltage source. This phenomenon, known as "electromagnetic induction," was discovered by Michael Faraday. It made our modern electrically driven society possible.

408 **Most electrical power is generated through the use of induction.** In an electrical generator, some source of energy (burning coal, falling water, etc.) is used to make a shaft rotate. The spinning shaft is attached to a loop of electrical wire between the poles of a magnet, and the fact that the loop is spinning in a magnetic field means an electrical current will be produced in it. This current is then tapped and run through power lines, eventually coming into your house to run your stove, your lights, your stereo system, and all the rest of your household appliances.

409 **Generators using electromagnetic induction usually produce alternating current.** It turns out that when a wire is rotated in the presence of a magnetic field, the current in the wire will flow one way for half the time and the other way the other half of the time. Since virtually all commercial electricity in the United States is produced by rotating generators, it is all in this form— what we call alternating current (AC).

Pop Quiz

How fast do the generators at your local power station turn? Answer: Sixty times per second—electrical current in the United States is rated at sixty Hertz, or one hundred and twenty changes of direction per second.

410 **You use transformers all the time, whether you know it or not.** A simple transformer is shown in the sketch. Electrical current runs into one loop of wire. This current sets up a magnetic field which in turn causes a current to flow in the second loop. Thus, one loop affects the other even though they do not actually touch.

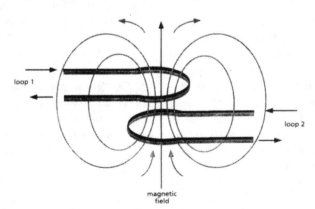

In a simple transformer, magnetic fields set up by currents in one loop create currents in the other.

The 115-volt current that comes out of your household wiring is normally stepped down by a transformer when it goes into your computer, stereo, TV set, or other electronic gadgets. You can recognize transformers because they tend to be rather bulky and heavy, with thousands of coils of wire wrapped around an iron frame.

411 For technical reasons, it is most economical to transfer electricity from generating plants to your home at very high voltage. Because of this, power lines leading away from major generating plants are typically 500,000 volts or more. These lines are carried on the large pylons you often see marching across the countryside. Before the electricity is fed into a power line, it is put into a transformer that increases the voltage and lowers the current. It is distributed around the city at 13,800 volts. Before it comes to your house, it goes through another transformer and is reduced to 230 volts. You can see these final stage transformers on power lines—they look like large metal garbage cans.

Maxwell's Equations

412 Maxwell's equations unify the phenomena of electricity and magnetism. They play the same role in electricity and magnetism that Newton's laws play in mechanics. First written down in the 1870s by the Scottish physicist James Clerk Maxwell, the four equations summarize everything we know about electricity and magnetism and the connections between them. They can be summarized as follows.

1. Unlike charges attract each other, like charges repel (Coulomb's law).
2. There are no isolated magnetic poles.
3. Electrical current gives rise to magnetic fields.
4. Changing magnetic fields can give rise to electrical current.

A Good Question

413 Why are these laws called Maxwell's equations when they were all discovered by someone else? Maxwell did not discover any of the four laws just stated. What he did was: (1) realize that these four laws were the very heart of the theory of electricity and magnetism; and (2) make a small addition to the third law just stated (he added something called the displacement current to the possible sources of the magnetic field). With this addition, the four laws fell into place and it became clear that everything we know about electricity and magnetism is contained within them.

One of the transcendent moments of my career as an undergraduate student was the day that my professor wrote Maxwell's equations on the blackboard and I realized that all the complex things that we had been studying for an entire year could be summarized in equations that could be written on the back of an envelope. From such moments are the careers of scientists made.

listen to a radio, or make a transatlantic telephone call, you are using technology that followed from Maxwell's work.

414 The most important immediate consequence of Maxwell's equations was his prediction of the existence of electromagnetic radiation, which led to the discovery and utilization of radio and microwaves. Every time you turn on a TV set,

415 Maxwell was a pioneer in color photographs. In fact, the first color photograph ever taken was part of his Ph.D. thesis. It is a surprisingly good photograph of a bunch of grapes. It can still be seen at Cambridge University, where he was a student.

Electromagnetic Radiation

416 Light, radio, and X rays are all examples of electromagnetic radiation. An electromagnetic wave is shown. You can think of the waves as electric and magnetic fields traveling through space the way water waves travel across the surface of a lake.

417 All electromagnetic waves travel at the speed of light. Electromagnetic waves travel through the vacuum at the speed of 186,000 miles (300,000 kilometers) per second. This is the speed of light, which is customarily represented as "c."

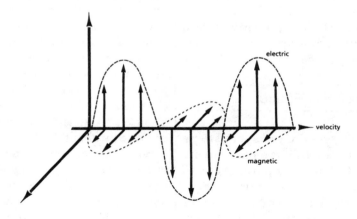

The electromagnetic wave.

418 James Clerk Maxwell first realized that such things as electromagnetic waves ought to exist. He discovered that the equations that now bear his name imply that there ought to be a wave that can travel through the vacuum (he would have said "through the ether"), and that these waves would have the structure shown in the diagram. When he calculated what the speed of these new waves should be, he found the prediction was 186,000 miles per second. He realized, of course, that this was the speed of light, and identified his predicted waves with light.

Once Maxwell realized that there was a new kind of wave implicit in his equations, he also realized that there should be many different kinds of these waves beyond ordinary light. He predicted the existence of things like radio waves, which were subsequently discovered in 1888 by Heinrich Hertz. Since that time, many other types of electromagnetic radiation have been discovered (see the following).

419 Most people associate the name of Guglielmo Marconi (1874–1937) with the discovery of radio. Marconi took Hertz's discovery and used it to send signals over long distances. It was he who sent the first radio message across the Atlantic.

420 All electromagnetic waves are contained in the electromagnetic spectrum. Today we know that there

are many kinds of electromagnetic waves, from radio (whose wavelengths are thousands of miles) down to so-called gamma rays (whose wavelengths are less than the size of an elementary particle). These waves differ from each other *only* in that they have different wavelengths—they are otherwise identical and have the structure shown. Here is a list of the electromagnetic waves that we know about:

Name of Wave	Wavelength
AM radio	tens to hundreds of miles
FM radio/TV	miles to feet
microwave	inches
infrared	thousandth of an inch
red light	8000 atoms
violet light	4000 atoms
ultraviolet	hundreds of atoms
X ray	a few atoms
gamma rays	size of an atom to the size of a nucleus

421 Visible light constitutes only a small part of the electromagnetic spectrum. Visible light is so important to human beings that we naturally assume that it must be very important in nature. In fact, it is only one of many different kinds of electromagnetic radiation, occupying a very small part of the electromagnetic spectrum—the part reserved for waves with a distance of four to eight thousand atomic diameters between crests. This is just another reminder of the relatively unimportant role that *Homo sapiens* plays in the grand scheme of things.

422 **The human body (but not the human eye) can detect other kinds of electromagnetic radiation.** When you hold out your hand to something that is warm, infrared radiation is carrying the energy from the object to your hand. Thus, the sensation of warmth is a "detection" of infrared radiation outside the visible range.

If you stay out in the sun and get sunburn, the cause of your discomfort is ultraviolet radiation. Thus, you can say that your body detects UV as well.

423 **When electrical charges accelerate, electromagnetic waves are produced.** Electrons being pushed through a wire generate radio waves and microwaves, electrons jumping in atoms generate light, and fast-moving electrons being stopped in a block of metal generate X rays. All electromagnetic radiation ultimately is produced by the motion of electrical charges.

424 **In a radio station's antenna, electrons are accelerated back and forth.** They produce the electromagnetic wave we call "radio." That wave travels through space to our radio antenna. In the antenna are more electrons which are, in turn, accelerated by that wave. The movement of these electrons constitutes an electrical current which, when fed through the circuits of your radio, is transformed into the sound that you hear.

425 **There are only two "windows" for electromagnetic radiation in the at-**

A crystal set, one of the earliest radios.

mosphere. Visible light travels easily through the atmosphere. When you are on a plane, for example, you can see a hundred miles through the air. In the same way, radio waves can travel long distances through the atmosphere. Have you ever picked up a radio station thousands of miles away from where you were driving your car? We say that the atmosphere has "windows" for visible light and radio waves.

All other forms of radiation are absorbed, either totally or partially, into the atmosphere. Astronomers wishing to detect these electromagnetic waves coming from distant objects usually have to go above the atmosphere. Satellite observatories have made extensive studies of X-ray, infrared, ultraviolet, and microwave radiation coming from space. In our generation, for the first time, we have been able to see everything that is out there, and this has produced a golden age of astronomy.

It has always seemed ironic to me that during the thousands of years that human beings have studied the stars, all this radiation was traveling millions of light-years to get to the earth, only to be absorbed in the last few miles by the atmosphere.

426 It is quite likely that within the lifetime of most people reading these words, we will have a "wrist radio"—a device that will convert your voice into a weak electromagnetic signal. This signal will then be picked up by an orbiting satellite antenna and sent back down (suitably amplified) to some other person's "wrist radio." Such a development, already prefigured by the cellular phone, requires only the ability to build detectors capable of picking up very weak signals.

Thus do comic book ideas become reality.

Note to reader: You may want to consult the sections on sensory systems, light, and quantum mechanics for further items that relate to electromagnetic radiation.

Magnetism

427 There is a magnetic force in nature. When your compass needle points north, or when you stick a note onto your refrigerator with a little magnet, you are using one of the fundamental forces in nature, the force we call magnetism. This force was known to all the ancient civilizations, including the Greeks and Chinese.

Anything capable of exerting a magnetic force (for example, by deflecting a compass needle) is a magnet. The most common magnetic material is iron, and there are many iron ores that have magnetic properties. Indeed, it was these naturally occurring magnets that led Greek scientists to investigate magnetism in the first place.

428 **The Greeks believed that there was an island in the Mediterranean made from naturally occurring magnetic materials.** They warned that ships should not be put together with nails, because if such a ship ever sailed near that island the nails would be pulled out and the ship would come apart.

There are, of course, good reasons to build a ship with pegs instead of nails, but the mythical "magnetic island" is not one of them.

429 **Magnets sometimes attract and sometimes repel each other.** We say that every magnet has two poles— we call them north and south—and that like poles repel each other and unlike poles attract. Thus, if you bring two north (or two south) poles of two magnets near each other, the magnets will be pushed apart. If, on the other hand, you bring the south pole of one magnet near the north pole of another, the magnets will be attracted.

430 **A compass needle is a magnet.** When we say that a compass needle points north, what we really mean is that a magnetic force acts on that compass needle. One end is attracted to the earth's north pole, the other to the earth's south pole. The net result is that no matter where a compass

needle points initially, it will be twisted around until it lines itself up in a north/south direction. This, of course, is why a compass is so useful in navigation.

Because the end of the compass labeled "N" points toward the earth's north pole, that end must actually be a south pole of the needle. To avoid confusion, physicists usually refer to the pole of a compass on which you paint the "N" as the "north-seeking pole."

431 **The earth is a magnet.** The fact that a compass needle responds to forces exerted on it by the earth proves that the earth is capable of exerting a magnetic force and is therefore a magnet. In fact, you can think of the earth as being very similar to a giant bar magnet.

432 **There are no isolated magnetic poles in nature.** As far as we can tell, every north magnetic pole that exists in nature is accompanied by a south magnetic pole. If you take an ordinary bar magnet and break it in two, you do not get a north pole and a south pole, but two short magnets, each of which has its own north and south pole.

Enduring Mystery

433 **Where are the magnetic monopoles?** A single isolated north or south magnetic pole would be called a "magnetic monopole." Physicists have expended a great deal of effort in searching for monopoles, but, with one disputed exception, they have not been successful. This is a great puzzle, because there are many isolated electrical charges in nature (for example, the electron and proton), and because we believe there is as profound symmetry between electricity and magnetism. To a physicist, living in a world with lots of electrical charges and no magnetic monopoles is something like looking at a huge painting which has a piece torn out of it—there is a constant nagging reminder that something is missing.

434 **All magnets are electromagnets.** Because there are no free magnetic monopoles in nature, all magnetic fields that we know about have to arise from the effects of moving electrical charges. For example, an electron in orbit around an atom constitutes an electrical current—a miniature, of course, but an electrical current just the same. It is this electrical current that can make the atom into a small magnet.

In the same way, we believe it is the motion of liquid iron in the earth's core that gives rise to the earth's magnetic field (see below) and that the motion of charged particles in the interior of the sun gives rise to the sun's magnetic field.

435 **The largest man-made magnetic fields are produced at the National Magnet Laboratory in Cambridge, Mass.,** and run to strengths 40,000 or more times that of the earth. The largest magnetic fields anywhere in the universe are probably those on the surface of pulsars, and may run as high as many billion times stronger than the earth's field.

436 **Natural magnets, the type that are usually made from iron, are called ferromagnets,** or, sometimes, "permanent magnets." Here's how a ferromagnet works: the "magnets" associated with each of the iron atoms tend to line up, as shown in the sketch. This lining up happens because of a force between neighboring atoms.

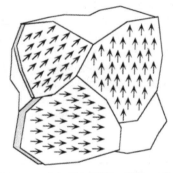

Ferromagnetic domains. When they reinforce each other, they create a permanent magnet.

The force that causes the atoms to line up creates what is called a "ferromagnetic domain." This is a block of material about a thousand atoms across in which all of the atomic magnets are pointing in the same direction. In a normal piece of iron, the domains point in random directions, so there is no net magnetic field outside the iron, even though there is one inside each domain.

When a piece of iron becomes magnetized, the domains are pulled around and made to point in the same direction. In this way, the magnetic fields of all the domains reinforce each other, and the material exerts a large magnetic force on anything near it. The more domains lined up in a given piece of material, the stronger its magnetic field will be.

437 **The existence of ferromagnetic domains explains why a magnet can be demagnetized.** A piece of iron or an alloy will be a magnet only so long as its domains are lined up. If a magnet is heated, however, the domains will be jostled around and will return to their normal random orientations. We say that the iron has been demagnetized. In order for it to be remagnetized, it has to be placed in a strong magnetic field so that the domains will once again line up.

438 **There are only a few naturally occurring magnetic materials.** Iron, of course, is the most common, but nickel and cobalt are also in this class. The most powerful permanent magnets are made from alloys of iron, boron, and neodymium.

The reason these materials and no others are magnetized is that the force that aligns neighboring atoms depends very sensitively on how far apart those atoms are. Only in these elements and some of their alloys is this distance just right for the formation of domains.

439 **Some materials are paramagnets.** These are materials that do not produce a magnetic field by themselves, but will do so if they are near another magnet. The way a paramagnet works is this: under normal circumstances, the atomic magnets are pointed in random directions and there is no magnetic field associated with the material. In the presence of another magnetic field, however, the atomic magnets in the material line up to reinforce and strengthen that external field. Examples of paramagnets are liquid oxygen and some ions of uranium.

440 **The rotation of the earth generates a magnetic field.** As the earth rotates, the liquid iron core rotates with it. Liquid iron will conduct electricity, although the fact that it has no net electrical charge means that it doesn't constitute an electric current in and of itself. There is, however, a complex process by which such a rotating conductor produces a magnetic field, and it is believed that this is the basic mechanism behind the earth's magnetic field. Thus, when you see a compass needle pointing north, you are dealing with a force whose origin is in the very heart of our planet.

441 **The earth's magnetic field undergoes episodic reversals.** Right now, the "north" pole of the earth's magnet is in the Canadian Arctic. There have been times in the past, however, when the north pole is where

Antarctica is now. We can document at least three hundred such reversals in the last few hundred million years. These reversals are erratic and the complete changeover of the poles seems to require about five thousand years. It seems to happen by having the field shrink down to zero and then grow again in the opposite direction, rather than having the north pole migrate across the face of the earth.

Enduring Mystery

442 **Why does the earth's magnetic field reverse?** It's not too hard to figure out how a planet could have a steady magnetic field. It is very difficult, however, to figure out how it could have a steady field that decides to change directions every once in a while. The erratic nature of the earth's magnetic field remains a profound mystery to geophysicists.

443 **Rocks remember where the earth's magnetic field used to be.** When molten rock comes to the surface of the earth, tiny bits of natural magnetic material are floating in it. These bits of material align themselves by pointing toward wherever the north pole happens to be at that time. Once the rock hardens, this direction is locked in, and the rock retains its memory of where the north pole was when it was formed. Thus, if we find such frozen "magnets" pointing south, we know that the "north" pole of the earth was at

what is now the south pole when the rock hardened. The study of old magnetic fields is called paleomagnetism, and this field provides some of the important evidence for plate tectonics.

444 **The sun, too, has a magnetic field.** The origin of the sun's magnetic field is probably similar to that of the earth's. The entire sun rotates and is made of a material that will conduct electricity—in this case, the plasma made of loose electrons and the atoms from which they have been torn. The sun's magnetic field seems to reverse itself every eleven years. As is the case with the earth, the origin of the sun's magnetic field (and the reasons for the reversals) is not very well understood.

445 **Sunspots are related to the sun's magnetic field.** The dark spots that are seen on the sun appear to be consequences of magnetic storms and magnetic phenomena under the sun's surface. The spots go through an eleven-year cycle, waxing and waning along with the magnetic field. They appear to be most numerous at the end of the cycle and least numerous in the beginning.

446 **Sunspots appear to us to be dark, but in fact they are incandescent.** They appear dark because even though they emit a lot of light, the material around them emits more. Our eye interprets this difference in illumination as a dark spot.

447 **Why has the sunspot cycle stopped in the past?** Although sun spots appear to go through a regular eleven-year cycle now, there have been several periods within recorded history when they stopped altogether. This, of course, adds to the mystery of the solar magnetic field. How can something that is ticking over like a clock now suddenly stop (often for periods of hundreds of years), then just start up again?

448 **The last time the sun spots stopped was between 1645 and 1715,** a period that coincides roughly with the reign of Louis XIV (called the Sun King) in France. Perhaps sunspots didn't dare to appear as blemishes during this august period.

Electricity

449 **Electrical charge is one of the fundamental properties of matter.** Like time, electrical charge is one of those things that's easy to point to and very difficult to define. We know that the electrical charge must be a basic property of matter because it is capable of generating forces. If you run a comb through your hair on a dry day and then bring the comb near a bit of paper, the bit of paper will move and stick to the comb. From Newton's first law, this means that whatever you did to the comb by rubbing it made it capable of exerting a force. We call this force "electricity," and we define an electrical charge as anything that is capable of producing the electrical force.

450 **There are two kinds of electrical charges:** like charges attract, unlike charges repel. The Greeks knew that if you rub a piece of amber with cat's fur or a piece of glass with silk, you get something capable of exerting an electrical force. They also knew that two pieces of amber would repel each other, but be attracted to the glass. This means that not only are there two kinds of electrical charges, but there are two kinds of electrical forces—attractive and repulsive. The former operate between unlike electrical charges, the latter between like electrical charges.

The names for the two kinds of charges were chosen arbitrarily to be positive and negative.

451 **The nature of the force between electrical charges is contained in Coulomb's law,** after French scientist Charles-Augustin de Coulomb (1736–1806). The law bears an uncanny resemblance to Newton's law of universal gravitation. It says that if there is a charge Q_1 and a charge Q_2, a distance R apart, then the force will be given by

$$F = \frac{KQ_1Q_2}{R^2}$$

where K is a universal constant analogous to G, Newton's constant of gravitation.

452 **In normal situations, only electrons move when objects acquire an electrical charge.** In uncharged materials, there are as many negative electrons as there are positive charges in the nuclei. When you rub a material, one of two things can happen. You can push electrons into it, in which case the material acquires an excess of electrons and we say that it is negatively charged. Alternatively, you can pull electrons out of it, in which case there is a deficit of electrons and we say that the material has a positive electrical charge.

453 **An electrical current is made of moving electrical charges,** usually (but not always) electrons. The most common electrical currents are those in your household wiring—the sorts of things that turn your lights on, run your radio, and cook your food.

When conductors are arranged in a continuous loop so that a current can flow around them unimpeded,

the conductors constitute an electrical circuit.

454 **Every electrical circuit has three parts.** There are three things necessary to make an electrical circuit work. They are: (1) a source of energy to make the electrical charges move; (2) an unbroken path on which the charges can move; and (3) a "load" or a place where the electrical energy is used. For example, when you plug in a light the source of electricity is the power company, the unbroken circuit is the copper leading through the lamp cord, and the load is the light bulb itself.

455 **The unit of electrical currents is the ampere,** often abbreviated as "amp" or written "A." It is named after André-Marie Ampère (1775–1836), a French scientist who was one of the pioneers in the study of electrical phenomena. One way to think about measuring current is to imagine a microscopic traffic engineer inside a wire, counting the number of electrons that go by a point. One ampere corresponds to 6 x 10^{18} electrons going by each second.

Here are some common items and the amount of currents that flow in them:

100 watt light bulb	—	1 amp
toaster	—	10 amps
TV set	—	3 amps
car battery	—	50 amps (while cranking)

Much larger and much smaller currents can be (and have been) produced, by man and by nature.

456 Voltage measures the "oomph" with which electrical charges are pushed through a material.

The unit of voltage is the volt, named after Alessandro Volta (1745–1827) the Italian scientist who built the first battery. Some common voltages are as follows:

flashlight battery	—	1.5 volts
car battery	—	12 volts
ordinary household receptacles	—	115 volts
heavy-duty household receptacle	—	230 volts

457 An electrical current generates heat.

When electrical charges flow through a material, they collide with the atoms that are already there. In the collisions the electrons give up a little of their energy and the atoms wind up moving a little faster than they were originally. We interpret this faster atomic motion as heat.

Except for superconductors, every material that carries electricity drains some of the energy away from the current and converts it into heat. You can feel this heat by touching the power cord of an electrical saw or an iron after it has been in use for a while.

The phenomenon by which a material converts some electrical energy to heat is called "resistance"— the more energy transferred to atoms, the higher the resistance of the conductor.

Pop Quiz

Is the wire in your toaster high or low resistance? Answer: It has to be high resistance because the toaster glows when electrical current runs through it.

458 The flow of electrons in a circuit is like the flow of water through a pipe.

If water flows from a city water tower through a pipe to your home, the various electrical analogues are as follows.

1. The height of the tower (which determines the "oomph" with which the water is being pushed through the pipes) corresponds to the voltage. The higher the tower, the higher the pressure on the water. In the same way, the greater the voltage in a circuit, the greater the flow of electrical current.

2. The quantity that corresponds to electrical current is the rate of flow of water. The more water flowing by a point each second, the higher the water "current."

3. The quantity corresponding to resistance is something like the size of the pipe. The smaller the pipe the harder it is to push high currents through it, and the more energy is expended in turbulence and heating in the water. In the same way, a high-resistance wire will impede the flow of electricity and cause much of the energy of the electrons to be converted into heat.

459 The power used by any household appli-

ance is measured in watts and is usually written on the little metal tag attached to the appliance. Some typical household power consumptions are:

stereo	—	200 watts
toaster	—	1 kilowatt
power saw	—	900 watts
TV set	—	300 watts

460 **The cost of running the appliance depends on the total energy used, and is measured (and paid for) in kilowatt hours.** For example, a TV set running for 6 hours uses 300 × 6 = 1800 watt-hours or 1.8 Kw-hrs of energy. A kilowatt hour of electricity in the United States costs 8 cents (on the average).

Pop Quiz

How much does it cost to run a stereo for 5 hours? Answer: 200 watts x 5 hours = 1000 watt hours = 1 kw-hr. The cost will be a total of about 8 cents.

461 **There are two kinds of electrical current in common use.** Large-scale commercial electricity comes from generators that produce a current in which electrons flow first in one direction in the wire, then the other. For obvious reasons, this sort of current is called AC (alternating current).

Current from a battery, however, flows in only one direction. This is called DC (direct current).

462 **The fact that you use AC at home doesn't mean that all the electrons run back and forth from the power plant to your light bulb 60 times a second.** In fact, because of all the collisions electrons make (see the foregoing), their net movement through a wire is slow—less than an inch per second. What happens is that all the electrons move in unison one way, then reverse, never getting very far from their original starting point.

463 **Electrical equipment is grounded.** Anyone who spends time around electrical equipment (and that means most of us) has to be aware that they may inadvertently become part of an electrical circuit—usually a circuit that flows through the person's body into the ground. To prevent this, electrical equipment is grounded—that is, it is built so that if an exposed part of the equipment comes into contact with a live voltage, a large current will immediately flow back through the circuit through a special wire. This current will trip the circuit breakers or burn out the fuse, disconnecting everything from the power lines and preventing anyone from being hurt. If a circuit keeps going out, *something is wrong*. Don't play games in this situation—get it fixed before you use the circuit again.

464 **A battery stores chemical energy, then converts it to electrical current.** Batteries consist of two dissimilar metals immersed in a material (either solid or liquid) called an electrolyte. In a car battery, for example, the metals are lead and lead oxide, and the electrolyte (battery acid) is a dilute solution of sulfuric acid. As the battery is discharged, chemical reations convert both the lead and lead oxide to lead sulfate (the white stuff that sometimes builds up around the contacts) and the electrolyte to water. A fully discharged battery, then, consists of two lead sulfate plates in water. At this point, there is no more stored chemical energy in the system. One step in this series of chemical reactions requires electrons to move from one plate of the battery to the other. They do this through the wires in the circuit to which the battery is attached, and it is these electrons that we perceive as the electrical current.

In a rechargeable battery, the chemical reactions can be reversed by running current through the battery in the reverse direction. Such a battery is only truly dead when enough impurities build up on the plates to stop further chemical reactions.

Time

465 **Scientists can't tell you what time is, only how to measure it.** There are two important questions you can ask about time. You can ask what it is, and you can ask how to measure it. The first question is the domain of philosophers, mystics, and others who like dealing with insoluble problems. Physicists only deal with how to measure time.

St. Augustine in his *Confessions*, said, "What is time? If no one asks me, I know what it is. If I wish to explain what it is to him who asks me, I do not know." This is probably as good a definition as you're likely to get.

466 **In order to measure time, you must have** **a regularly recurring phenomenon in nature.** The standard technique is to find something that happens regularly, and then define the unit of time in terms of the reappearance and recurrence of the phenomenon. For example, one unit of time is the "day"—the time between two successive sunrises. All systems for measuring time depend, ultimately, on the recurring phenomenon that is chosen to define the basic standard.

Throughout most of human history the passage of time has been measured in terms of the day (which is related to the time that it takes the earth to turn once on its axis) and the year (the time it takes the earth to go once in its orbit around the sun).

467 The Egyptians defined the hour to be one-twelfth of the time between sunrise and sunset. This meant that for the Egyptians the length of the hour was different from one day to the next, and was not the same during the day as it was at night.

468 The first exercise in the measurement of time was the production of the calendar. When human beings began to develop agriculture, it became necessary for them to mark important events like the planting time for particular crops. In other words, they had to have a calendar. The calendar is really a clock that "ticks" once a year and therefore keeps track of where the earth is in its orbit around the sun. It is this position that determines the seasons.

The basic problem in constructing a calendar is that the number of days in a year is not an even number. The following calendars represent successive approximations to the true length of the year:

469 Egyptian Calendar. This calendar consisted of twelve months of thirty days each, followed by a five-day party. The problem with the Egyptian calendar arose from the fact that there are approximately 365 ¼ days in a year, not 365. This meant that the calendar would "slip" a quarter day every year. These slippages built up, and, if you had followed things blindly, would eventually have led to a situation where you had the Egyptian equivalent of snow in "August."

470 Our modern New Year's Eve parties trace their way back to the Egyptian end-of-year bash. It was a time that didn't really belong in the year, hence a time when nothing really counted. Anything went. We may have a more modern calendar these days, but we seem to have managed to retain the truly important part of this old one.

471 Julian Calendar. The calendar introduced by Julius Caesar tried to bring some order into time keeping in the Roman Empire. It solved the problem of the extra quarter day by introducing the leap year. Every four years the year is one day longer, and this makes up for most of the slippage that appeared in the Egyptian calendar. It didn't catch all of it, though, because the year is 11 minutes 14 seconds shorter than 365 ¼ days. These errors started to accumulate (they amount to 7 days every 1000 years) until they began to mess up the observance of Easter. This led to the . . .

472 Gregorian Calendar. The Gregorian calendar was introduced by Pope Gregory in 1582 to deal with the accumulated slippage in the Julian calendar. It works by dropping leap years when they fall on centennials except when the centennial is divisible by four. Thus, 2000 will retain its leap year while 1700, 1800, and 1900 did not. The Gregorian calendar is the one we use today and the one with which you are familiar.

473 **Russia didn't adopt the Gregorian calendar until after the Revolution.** Thus, for several centuries, there were two calendars operating in Europe—the Gregorian in most of the West, the Julian in the East. This explains why you often see dates in Russian history given twice—one in modern (Gregorian) terms, the other in "old style" (Julian) terms.

474 **46 B.C. was the longest year on record, 1582 the shortest.** When Julius Caesar introduced his calendar in 46 B.C., he added two extra months to the year as well as 23 extra days to February to make up for the accumulated slippage in the Egyptian calendar. Thus, 46 B.C. was 455 days long, the longest year on record.

Similarly, when Pope Gregory introduced his calendar in 1582, he decreed that the day Oct 5 would be Oct 15, making it the shortest year on record.

475 **The rotation of the earth is actually a very poor time standard.** If you look at the rotation closely enough, it is quite unsteady. The gravitational pull of the moon and the planets, the effects of the tides, earthquakes, and even the motion of the winds cause the rotation to slow down and speed up erratically. These changes aren't huge—they're on the order of milliseconds per day, but if you define the second to be a particular fraction of the length of a day (which is the way it used to be done), the second will change from one year to the next.

476 **The second is now defined in terms of the motion of an electron in an atom.** In 1967, the International Commission on Weights and Measures redefined the second in terms of the time it takes for an electron to spin on its own axis inside an atom of cesium. This is the standard that is used today.

The so-called atomic clock can measure the length of a second to an accuracy of thirteen decimal places.

477 **We now keep our calendars accurate by inserting a leap second into the year.** Here's how it works: there are a number of clocks at various national laboratories around the world. When a majority of the clocks agree that the rotation of the earth has gotten out of line by half a second, a "leap second" is inserted into a chosen day at midnight. This is done every few years, most recently on Dec 31, 1990.

478 **The most accurate clock is the hydrogen maser.** Although the cesium-based atomic clock is the best time standard at the present time, another regular motion, the movement of an electron in a hydrogen molecule, can produce a clock that is significantly more accurate. The cesium clock is accurate to thirteen decimal places, the hydrogen to fifteen. Unfortunately, the hydrogen maser clock is stable for less than a second, so it can't be used as a long term time standard.

479 The longest time that anyone has ever tried to measure is the lifetime of the proton—more than 10^{33} years.
The longest time anyone has actually measured is the lifetime of the universe—about 16 billion years.
The shortest time that has been measured (indirectly) is the decay of some elementary particles—10^{-24} seconds.
The shortest time that can be measured directly are bursts of light in special lasers—10^{-15} seconds.

480 The nanosecond is the relevant time scale for modern fast electronic systems. Light travels about one foot in a nanosecond, or a billionth of a second, and signals in an electrical circuit can't travel any faster. This means that a computer that is much bigger than a few feet across

will not be able to move signals from one side of its body to another in times less than a nanosecond. This is considered to be a fundamental limitation on the possible speed of computers.

481 The time it takes a neuron to fire in your nervous system is about a millisecond, whereas the fastest switch in modern computers can close in a picosecond. The fact that your brain can do many things (such as process visual information) faster than even the largest computer has to do with the superiority of the design of the circuits rather than with the speed of the individual components.

The names attached to various short times are:

Name	Time
millisecond	.001 sec (thousandth)
microsecond	.000001 sec (millionth)
nanosecond	.000000001 sec (billionth)
picosecond	.000000000001 sec (trillionth)
femtosecond	.000000000000001 sec (million billionth)

Gravity

482 The first modern theory of gravitation was propounded by Isaac Newton. He called it the the law of universal gravitation. The law states that

every object in the universe exerts an attractive force on every other object in the universe, and that this force depends on how massive the objects are and how far apart they are. The more massive, the greater the force; the greater the separation, the smaller the force. In equation form, the law looks like this:

$$F = \frac{G\ M_1\ M_2}{D^2}$$

where F is the attractive force of gravity, M_1 and M_2 are the masses of two objects, D is the separation between them, and G is a number known as Newton's constant of universal gravitation.

Enduring Mystery

483 **Did Newton really see the apple?** Newton's account of his discovery of the law of universal gravitation has entered the folklore along with George Washington's cherry tree and Ben Franklin's kite. According to Newton, here's what happened: he was walking in an orchard one day when he saw an apple fall from a tree. At the same time, he saw the moon in the sky. He realized that if the force that made the apple fall extended as far as the moon, that it might explain why the moon stayed in its orbit.

Historians question whether Newton really saw the apple or whether this story was cooked up later to establish his priority in explaining the orbit of the moon.

484 **Newton's theory of gravity was the first unified field theory.** Before Newton, it was assumed that the force that made things fall to the surface of the earth was not the same as the force that moved the sun, moon, planets, and stars. What Newton did was show that there is only one kind of gravity, thereby unifying the two seemingly separate forces.

485 **When an apple falls to the earth, it is easy to overlook the fact that the earth is also being pulled up toward the apple.** Newton's law of gravitation allows you to calculate how much the earth moves while the apple is falling—it turns out that the earth will move up to meet the apple by a distance less than the size of a single atomic nucleus. Needless to say, there's no way you could ever measure such a movement of our planet.

486 **According to Newton, EVERYTHING exerts a force on EVERYTHING else.** As you read this, the earth is pulling down on you—that's what explains why you don't float off your chair. In fact, the law says that every object in the universe is pulling on you at this moment. In addition to the earth, the building in which you are sitting, the tree outside your window, and distant stars are all exerting gravitational forces on you, and you are exerting forces on them as well.

Of course, for all practical purposes, you can neglect all of those forces except for that of the earth, since it is so much stronger than the others. It is not unheard of, however, for physicists who engage in very precise measurements to have to take into account the gravitational force exerted on their apparatus by the steel and concrete in the buildings in which they are housed.

487 **Astrology has no scientific basis.** Astrologers sometimes use the universal aspect of gravitation to argue that it is possible, at least in principle, for stars and planets to have an influence on children at birth. However, you must remember that *everything* is exerting a gravitational force on that child, including the attending physician and nurses in the delivery room. When you put in the numbers, you find out that the attending physician exerts a greater gravitational force on the child than does Alpha Centauri, the nearest star.

488 **Ocean tides are caused by the force of gravity,** but the explanation of ocean tides is a rather complicated business—it's not just a simple matter of the moon pulling on the water in the ocean. As evidence for this statement, consider: there are two tides a day (instead of one, which you might expect on the basis of a simple picture). In addition, high tide usually occurs when the moon is on the horizon, rather than directly overhead (as you might expect if you thought tides were caused by the water rising toward the moon).

There are two tides because, while the moon pulls the water away from the earth on the side facing it, it pulls the earth away from the water on the opposite side. The net result, as seen by someone on the earth, is that the water is raised above the normal surface on two diametrically opposite sides of the earth. The fact that the tides occur when the moon is on the horizon, rather than overhead, has to do with the fact that the oceans are relatively shallow, so that the tidal bulge cannot travel through the water fast enough to keep up with the motion of the moon.

489 **There are land tides as well as ocean tides.** Although tides in the ocean are the most dramatic evidence we have of gravitation, there are also tides on the land. Typically, as the moon passes overhead, the level of the "solid" earth will go up a few inches and then go down again. Since this motion takes place over a twelve-hour period, it usually isn't noticed. Every solid object in the universe—from the earth to the moons of Saturn and beyond—shows land tides if it is located near a large object. Ocean tides, on the other hand, exist only on the earth.

490 **The sun contributes to tides in the earth's oceans,** tides that are about a third as high as those due to the moon. This is why the high tides at certain times of the month are higher than they are at other times. During the new moon and full moon, when the tides of the moon and the sun reinforce each other, we have the so-called "spring tides." At the quarter moons, the two are out of step and we have lower high tides—the so-called "neap tides."

491 The effect of tides keeps the same face of the moon pointing toward the earth. The law of universal gravitation tells us that the earth must raise land tides on the moon. It turns out that when you calculate the effects of land tides on the moon, you find that after a few hundred thousand years the satellite (moon) winds up keeping one face toward its larger partner. In the jargon of astronomers, we say the satellite has been "despun." All large moons in the solar system have been despun.

Tides in a town in Nova Scotia, showing the dramatic change in sea level due to the moon's gravitational attraction.

492 **General relativity is our best modern theory of gravitation.** The central tenet of the theory is that the presence of matter warps the fabric of the universe. An easy way to visualize how general relativity works is to imagine stretching a garbage bag tightly over the head of a waste basket, then placing a ball bearing on it so that the ball bearing distorts the surface. If you then roll a little marble along the plastic, it will be deflected from its course. If you didn't know about the warping of the plastic, you would say that the ball bearing was exerting a force on the marble.

In Newtonian gravity you say that one object exerts a force on the other. In general relativity, on the other hand, you say that one object distorts the fabric of space-time and therefore causes changes in the motion of other objects.

493 **Einstein's theory does not overthrow Newtonian gravity.** If you look at the equations of general relativity and extrapolate them to a regime where masses are small, you find that they become exactly the equations you would get from Newton's law of universal gravitation. In other words, general relativity *contains* Newtonian gravity and extends it, but does not invalidate it for the region for which it was intended to be used.

494 **There are very few places in the universe where general relativity is im-**portant. For everyday life, for sending space probes around the solar system, and for any other situation you are likely to encounter, you don't need to worry about general relativity. The reason is that the effects of general relativity are normally so small as to be completely negligible, and the good old Newtonian theory of gravity works just fine. You have to go to general relativity only in cases where objects get very massive (e.g., near black holes), when very large distances are involved (e.g., when doing cosmology), or when taking very precise measurements. In all other cases, you can ignore the fact that Newton's law is only an approximation to a better theory.

495 **General relativity predicts that light will be bent when it comes around the sun.** In 1919, Arthur (later Sir Arthur) Eddington made what most people regard as the most dramatic and best-known confirmation of the theory of relativity. Traveling to the coast of Africa to observe a solar eclipse, he noted that the apparent position of stars near the edge of the sun appeared to be shifted during the eclipse, a shift that could only occur if light from the star were bent as it went around the sun.

Both Newtonian gravity and general relativity predict that such a bending should occur, but they predict different numbers for the angle of deflection. When Eddington verified the predictions of Einstein's relativity, it caused a sensation in the press and started Einstein on the path to becoming a major public figure.

Today, "light" bending measurements are done on radio waves from distant quasars, rather than with light. They confirm general relativity to an accuracy of about 1 percent.

496 **There are two other tests that make up the classical verifications in the prediction of general relativity.** One of them has to do with a very small effect on the orbit of the planet Mercury—an effect called the "advance of the perihelion." If you look at the elliptical orbit of the planet over long periods of time, the axis of the ellipse rotates slowly. In other words, the point at which the planet makes its closest approach to the sun moves steadily over long periods of time. The advance isn't much, less than one degree per century. Most of this so-called "advance of the perihelion" is due to the gravitational effects of the other planets, particularly Jupiter. A small amount, however—some forty-two seconds of arc per century—was unexplained until Einstein showed that it was due to the effects of general relativity.

Starting in the 1960s, astronomers have used radar ranging to make very accurate observations of the orbits of all the inner planets, from Mercury to Mars. These determinations of the advance of the perihelion of the planets are probably the most stringent tests of general relativity available at the present time.

The other test of relativity was concerned with the fact that as light moves up the surface of the earth, the force of gravity acts upon it and causes a lengthening of its wavelength. This effect was verified in the late 1950s.

497 **General relativity was accepted on very thin experimental grounds**—only three experiments. I suspect this is because the theory is so beautiful that scientists *want* to believe it. Nevertheless, many scientists are now engaged in devising new tests for the theory using modern high precision instruments.

498 **There will be tests of general relativity in the future.** Sometime in the mid-1990s a satellite will be put into orbit with a small sphere inside of it. General relativity predicts that there will be small deviations in the rotation of such a sphere because of the rotation of the earth. By measuring the rotation of the sphere to an as yet unheard-of precision, scientists will then be able to add another test of the theory to their repertoire.

In another test, scientists on earth are using very fast electronics to measure the time it takes light signals to travel east and travel west. The general theory of relativity says that there should be a small difference between these two times due to the rotation of the earth. Verification of these predictions is in the offing.

I expect that over the next decade,

modern advances in instrumentation will make it possible to detect many more small effects predicted by general relativity, and therefore to provide us, for the first time, with a firm experimental basis for our belief in the theory.

499 The spinning quartz ball to be used in the test of general relativity is the world's roundest object. Only a few inches across, if it were blown up to the size of the earth the highest "mountain" would be only a foot high!

500 General relativity cannot be the final theory of gravitation. The reason is simple. There is no room in the theory for quantum effects. This means that if you look at gravitational effects on a very small scale (on distances much smaller than the size of a proton, for example) the theory must break down and be replaced by something else. Presumably this new theory will contain general relativity as a special case, just as general relativity contains Newtonian gravity.

The attempt to reconcile the two great ideas of twentieth-century physics—quantum mechanics and relativity—occupies many of the best minds of this generation of theoretical physicists. I wish I could tell you that a great deal of progress is being made in the area, but I have a sneaking feeling that this is a task we're going to leave to the next generation.

501 One of the best of the modern "quantum graviteers" is Stephen Hawking, whose book *A Brief History of Time* was a best-seller in 1989, and whose affliction with ALS (Lou Gehrig's disease) makes him a larger-than-life figure in our society. Hawking approaches the problem of quantum gravity by "grafting" the two theories together. In other words, he puts relativity and quantum mechanics together by hand, rather than (as other physicists try to do) by developing a general principle from which the wedding of the two theories will be a natural outcome. One consequence of Hawking's work is his prediction that black holes, which are absolutely stable in the general theory of relativity, will convert their mass into radiation (so-called Hawking radiation). The black holes will then disappear after long periods of time. His ideas are also behind the new field of "baby universes" (see the following).

502 The place where quantum mechanics and gravity come together most dramatically is in studies of the early universe. The phenomenon of inflation, for example, is an effect that involves both the interaction of elementary particles and the warping of space-time by matter—i.e., it is a hybrid of quantum mechanics and general relativity. In addition, Stephen Hawking's "baby universes" and Alan Guth's "universe in your basement" are both ideas that depend on the coming together of quantum mechanics and gravity.

In both of these theories, you imagine the space-time fabric of the universe as being similar to a large balloon which, on a very small scale, has bumps and irregularities in it. These bumps and irregularities are caused by quantum effects. In some theories it occasionally happens that one of these irregularities can grow, like an aneurysm. If it does so, it can form its own little universe that travels along parallel to ours (hence the name "baby universes").

In Hawking's view, such universes are being spawned all the time. Guth, however, asks if it would ever be possible, even in principle, for human beings to create a universe by producing and manipulating the aneurisms. He calls this "making a universe in your basement." His answer as to whether it's possible: probably, but not anytime soon.

Classical Mechanics

503 Isaac Newton (1642–1727) can rightly be called the father of modern science. With his work, the last vestiges of medieval science were finally left behind, and the enterprise of modern science began in earnest.

The essential feature of the Newtonian vision of the universe can be best understood by picturing a mechanical clock whose gears kept turning according to well-defined laws. Newton felt that the solar system and the rest of the universe were ticking along according to laws that human beings could both discover and understand. And, just as a clock will continue ticking once it has been wound, the Newtonian universe was thought to have been wound once by God and to be traveling along according to his designs, but without his direct intervention.

504 Newton's most productive year was from 1665 to 1666. The bubonic plague (or Black Death) was prevalent in England, and caused Cambridge University to shut down for a period of eighteen months. Isaac Newton, a student at the university, returned to his family's farm and for this entire period had no one to talk to about science. With only his own thoughts to guide him, he invented calculus, discovered the law of universal gravitation, and made a number of other lesser discoveries. It is hard to think of a more productive period in science, and the fact that it was the result of one man working alone makes it all the more amazing.

505 The heart of Newtonian mechanics is Newton's three laws of motion. These laws can be stated roughly as:

1. Nothing happens unless a force acts.

2. The force applied to a body is equal to its mass times its acceleration ($F = ma$).

3. For every action there is an equal and opposite reaction.

That's it! On the basis of these simple laws, together with the law of universal gravitation, Newton laid the foundation for modern science.

506 The work that Newton did is usually called classical mechanics. Newton concerned himself almost exclusively with the motion of material objects. The quintessential Newtonian problem is a collision of two billiard balls on a pool table. Newton showed that his laws allow you to predict the final positions and velocities of material objects, given their initial positions and velocities. Thus, the laws give you a way of following a system, once you know where it is at any given time. "Mechanics" is an old-fashioned term for the study of motion, and modern physicists add the prefix "classical" to distinguish Newton's work from quantum mechanics.

507 Perhaps the most ambitious statement of the Newtonian worldview was made by the French mathematician Pierre-Simon Laplace (1749–1827), who suggested a thought experiment involving the "divine calculator." The basic thesis is simple: if I knew the position and velocity of every particle in the uni-

verse at any given time, then I could use Newton's laws to predict the position and velocity of those particles at any time in the future. Of course, it would take someone of godlike prowess to make the calculation (hence the term "divine calculator") but in principle the calculation could be done. Laplace then pointed out that if one of those particles happened to be an atom in your hand (for example), then the divine calculator could tell you where your hand would be at any point in the future. Do you see the implications for all those sophomore bull sessions on the question of free will and predestination?

508 Newton's first law embodies the principle of causality. It says that an object will continue moving as it is moving unless a force acts on it—that nothing happens without a cause.

Newton's First Law can be thought of as a diagnostic tool. If we see a change in motion, we know that a force must have acted. Thus, when we encounter new situations (the motion of one magnet near another, for example) the laws give us a framework in terms of which to approach problems, even if those problems don't involve billiard balls.

509 Just because a force acts, it does not follow that there must be motion. For example, you can put your hands together in front of you and push. You are exerting forces—you can feel them—but nothing moves. This is because the force exerted by one hand is cancelled by the other. Lack of motion does not always signify the total absence of forces, then, but can also result from the presence of forces that cancel each other.

510 **Newton's second law relates acceleration to force.** The harder you push on something, it says, the faster it will go. The more massive it is, on the other hand, the harder you have to push. These commonsense statements square with our experience. The second law simply tells you how much faster something will go, how much harder you have to push.

511 **Newton's second law allows you to predict subsequent motion given present motion.** Doing so usually involves several steps. First, you have to identify the forces that are acting. Then you incorporate these into the equation form of Newton's second law. Finally, you solve the equation. Every professional physicist has spent a good portion of his or her youth carrying out these kinds of operations for successively more complex physical situations and problems.

512 **Newton's third law explains how a rocket can move in space even though there's no air for it to push against.** Think about firing a rifle. You pull the trigger and a force is exerted on a bullet, pushing it out the front end. At the same time, the third law says that an equal and opposite force is exerted on the rifle.

This is the force you feel as the recoil against your shoulder.

An engine in a rocket or jet aircraft exerts a force on hot gases and pushes them backwards. An equal and opposite force then pushes forward on the craft, providing it with propulsion.

513 **In practice, even Newtonian systems become unpredictable very quickly.** Although it is possible in principle to use Newton's laws to calculate the future positions and velocities of a number of moving billiard balls, if there are more than a few balls on the table it is impossible for a computer to carry calculations more than a few tens of collisions into the future. After that, the inevitable rounding errors in the computer and the inevitable measurement errors in defining the initial position and velocity of the balls blur things to the point that the predictions are no longer reliable.

514 **Galileo discovered the law that governs the motion of objects near the earth's surface.** He discovered the laws of falling bodies and showed that an object falling near the earth will be accelerated as it moves downward, and then determined the rate of acceleration. (For the record, the increase in velocity of a falling body is 32 feet per second during each second of fall.)

515 **Isaac Newton provided scientists with a game plan that they could use**

when exploring new areas—a plan we call the scientific method. In essence, the scientific method contains two steps. First, scientists observe nature and summarize the results of these observations in mathematical structures called theories. Next, the theories are used to predict the results of new, as yet unobserved processes, and these predictions are compared to what happens when the new experiments of observations are carried out. In this way, knowledge is extended to new areas.

Lest I be accused of oversimplifying what is, after all, a complex human endeavor, I should add that observation and the construction of theories are not separate activities, but go hand in hand and influence each other.

Physical Properties of Matter

516 The way any material behaves depends on how the atoms in that material are locked together. *Every* property of materials is related to atoms, and to make this point I list a variety of properties and relate them to atoms.

517 Pressure. When you inflate a tire, you are pumping lots of air molecules into it. These molecules rattle around inside the tire, and when they hit the tire wall, they rebound. Each rebound exerts a tiny force on the tire, and the pressure you read on your tire gauge is just the sum total of all those forces.

518 Air and water pressure. Both air and water are made of molecules, and both are therefore capable of exerting pressure. Molecules in a cube of water in the middle of the ocean, for example, will exert a pressure on all sides of the cube—up, down, and sideways. If you imagine a column of water extending down into the ocean, the force of gravity downward on that column has to be balanced by the upward force of pressure exerted by the water below it. Thus, the deeper you go into the ocean (or the atmosphere), the higher the pressure. At sea level, for example, the air exerts a pressure of 14.7 pounds per square inch.

519 Buoyancy. If something is put into water, a pressure will be exerted on it. The result of this pressure is an upward force we call buoyancy. The buoyant force is equal to the weight of the water displaced by the object, so if the object is less dense than the water it will float. Otherwise, it sinks.

Pop Quiz

How come an ocean liner floats when iron is heavier than water? Answer: The amount of water displaced by the ship is equal to the volume of iron *plus air* in the hull. If the ship were full of water (or iron), it would sink.

520 **Archimedes is supposed to have discovered the law of buoyancy when he was asked to determine if a crown was made of pure gold or had been adulterated.** Getting into his bath one day and noting the water rise as he did so, he is supposed to have shouted *"Eureka!"* ("I have found it!") and run down the street to try out his idea. The notion: by measuring the water displaced by the crown and then the water displaced by an equal weight of gold, he could tell whether the crown was pure gold or not.

History does not give a clear verdict about whether the crown was gold, nor does it tell us how bystanders reacted to Archimedes' discovery.

521 **Adhesion and Cohesion.** When the molecules of a material are attracted to other molecules of the same material, we call the force cohesion. This force holds things together. When molecules of different materials are attracted to each other, the force is called adhesion. This force makes one thing stick to another. In both cases, however, the basis for the force is the attraction between atoms.

522 **Surface tension.** Cohesive forces within a liquid tend to pull the liquid into a ball. When a drop of water "beads up" on your raincoat, it is the cohesive force that is pulling it together. Physicists think of the effects of cohesion as a force pulling the surface together, and call that force surface tension.

523 **Elasticity.** This is the property of solids that makes them return to their original shape when they are deformed. When you bend a piece of metal, the atoms in it exert a force that opposes the bending. As soon as you let go, the inner forces take over and the metal snaps back.

524 **Compressibility.** Because the forces between atoms can become repulsive if the atoms are pushed too close together, materials resist outside forces that try to compress them. Certain materials, like steel and water, resist very strongly. Others, like air, do not.

525 **Tensile strength.** Just as materials resist having their atoms pushed together, they resist having them torn apart. The tensile strength measures the force required to overcome the attractive forces between atoms and pull a material apart. Steel has a high tensile strength, while paper does not. Oddly enough, glass has a high tensile strength as well—it's hard to pull it apart, although it's easy to break it.

526 **Osmosis.** If two solutions are separated by a membrane, water (but not the molecules in solution) can move through the membrane, changing the concentration of the solution of both sides. This is called osmosis. When your skin is puckered after you've been in the bathtub a long time, it's because water has flowed into your cells by osmosis.

527 **Diffusion.** When molecules of two different fluids or gases come together, the normal molecular motion results in the two sets of molecules intermingling. This process is called diffusion. If you put a drop of ink into a glass of water, you can trace the progress of diffusion as the ink spreads.

Since diffusion depends only on the motion of molecules, it can crop up in unexpected places. It is well known to engineers, for example, that gases can diffuse into (and even through) metal containers. Space scientists have to worry about gases diffusing out of the walls of a spacecraft on long missions.

528 **Capillarity.** If you put a thin hollow tube down into a fluid, the fluid will rise in the tube. This effect is called capillarity. The way it works is this: the downward pull of gravity on the fluid in the tube is overcome by the force of adhesion (see the foregoing) between the fluid and the walls of the tube. It is capillarity that lifts water up in plants.

For a given-size tube, there is a limit to how high the fluid can rise. The weight of the fluid column cannot exceed the upward force exerted by cohesion.

Thermodynamics, Energy, and Heat

529 The study of heat began with the invention of the steam engine at the beginning of the Industrial Revolution. The need to have a better source of energy than competitors drove English, French, and German scientists to develop what we today call thermodynamics. It was a case where technology drove basic research rather than vice versa.

Today, thermodynamics is the science that tells us about the behavior of things like energy, waste heat, and the efficient use of resources. It also has developed into the science that gives us one of the best insights into the relation between the large-scale world we live in and the world of the atom.

Energy and Power

530 The concept of energy is one of the

most important ideas in thermodynamics—indeed, in all of science. Although the term "energy" has many colloquial meanings, to a physicist it means just one thing. A system has energy if it is capable of exerting a force over a distance (or, in the jargon of physicists, if it is capable of doing "work").

There are different kinds of energy. If an object is moving, it is capable of exerting a force on anything it runs into. Thus, it possesses energy. Energy of motion is called kinetic energy.

Similarly, an object can have energy by virtue of its position. If you lift this book, it will fall if released. During its fall, it is capable of exerting a force, and hence, of doing work. Energy associated with position is called potential energy.

Finally, an object can have energy by virtue of its mass ($E = mc^2$ and all that).

531 **There are many different kinds of potential energy.** If you think about lifting this book, you realize that the book has energy because the force of gravity acts on it to pull it down. We therefore call the energy of an elevated object "gravitational potential energy."

In the same way, an electron near the nucleus of an atom has energy because it could fall to a lower orbit. In this case, the fall would be occasioned by the electrical force, so we refer to it as "electrical potential energy."

When the electrons of molecules rearrange themselves during chemical reactions, they are changing

Water going over Niagara Falls is an example of gravitational potential energy being transformed into other kinds of energy, such as kinetic energy, sound, and seismic waves. If you stand near the bottom of the falls, you can feel the energy pouring through as the ground shakes.

their electrical potential energy. Energy stored in the arrangement of electrons is called "chemical potential energy."

There is also potential energy associated with magnetism, the forces that give rise to elasticity in solids, and with other forces.

532 **What we call "heat" is a form of kinetic energy at the atomic level.** When an object is hot, the atoms in it are moving around very rapidly. When it is cold, on the other hand, the atoms move slowly. Thus, what we call heat is actually a form of energy of the motion of atoms. The realization that heat could be explained in this way was one of the profound insights of nineteenth-century physics.

It also provides us with one link between the macroscopic world of our senses and the invisible world of the atom. If you can translate the sight of a red-hot burner on a stove and the feel of heat emanating from it into a vision of rapidly moving atoms, you're well on your way to the modern vision of the physical world.

533 **The forms of energy are not fixed immutably—one form can readily be exchanged for another.** For example, when you rub your hands together in cold weather, the kinetic energy of your hands is being converted into heat. Conversely, when you boil water over a campfire, the chemical energy of the wood is converted into heat in the water, and that heat is converted into energy of motion in the jet of steam that you see coming out of the kettle.

534 **The most important single fact about energy is that the total amount of energy in an isolated system does not change.** In the language of physicists, we say that energy is "conserved" and we talk about the law of "conservation" of energy.

The law of the conservation of energy is simply a restatement of a piece of folk wisdom—to wit, that there is no such thing as a free lunch. If you want to warm your house, you must get the energy to do so from somewhere—be it an electrical power plant or an oil fire. Energy cannot be created; neither can it be destroyed. All we can do with energy is exchange one form for another.

535 **The first law of thermodynamics says that heat is a form of energy, and that energy is conserved.** This is arguably one of the most important laws of nature ever discovered.

536 **"Power" refers to the rate at which energy is expended.** If you walk up a flight of stairs slowly and then run up those same stairs as fast as you can, you expend the same amount of energy each time. When you run, however, your power output is greater because you are expending the energy more rapidly. That's why you're breathing hard.

537 **Virtually all the energy used on earth comes from the sun.** It is converted from energy in the form of light to chemical energy by the process of photosynthesis. It can then be stored (in coal or oil, for example), used directly by plants, or, secondarily, used by animals. The energy you are using to hold this book up and move your eyes as you read comes, ultimately, from the sun by this route. Eventually, the energy we use from the sun becomes waste heat (see the following) and returns to space as infrared radiation.

Thus, we do not consume energy so much as use it as it passes through our environment.

538 **Power is measured in kilowatts or horsepower.** The unit of power in the metric system is the watt. The kilowatt, as the name implies, is a thousand watts. It might take several hundred watts to run a stereo.

The unit of power in the English system is the horsepower. The motor in an electrical table saw might develop one horsepower, the motor of a car several hundred.

539 **The unit of energy most familiar to us is the kilowatt-hour, which is reported on our electrical bills.** This is the amount of energy expended by a one-kilowatt source operating for one hour. A kilowatt hour will buy you:

one hour of a small electric bar heater.
ten hours on a light bulb.
four hours of TV.

540 **James Watt (1736–1819), the Scottish engineer who invented the modern steam engine, also defined horsepower.** He needed a way to sell his engines to mining engineers. At the time, water was kept out of mines by pumps driven by horses. Watt measured the rate at which a typical horse could do work over extended periods of time and then rated his engines accordingly. Thus, he could tell a prospective client that a one-horsepower engine would replace one horse in a mine.

Heat

541 **Although energy is conserved, it need not remain in the same place all the time.** This is particularly true of heat, which moves quite easily from one place to another. If you turn off the furnace in your house during the winter, for example, the house quickly gets cold as the heat moves outside. There are three processes by which heat can be moved from one place to another—conduction, convection, and radiation.

542 **Conduction.** If the inside of your house is warm, the molecules on the inside of the wall will be moving faster than the molecules on the outside. When atoms from the inside collide with those near the outside, energy will be transferred and the atoms near the outside will start to move faster. Thus, after a period of time, the outside of the wall will be warmer than it would be ordinarily. We think of this process as a "heat flow" from the inside to the outside of the house but, in fact, it is simply a matter of energy being transferred down a chain of atoms by ordinary collisions.

543 **Convection.** On a hot summer day, air over the land is warmed. Because it is warmed, it expands and becomes less dense than the colder air above it. Eventually, the situation becomes unstable. The warm air moves up and is replaced by cold air moving down. As the warm air moves up, it carries heat with it in a process called convection.

Convection is actually a much more efficient means of transferring energy than conduction. Convection is what transfers heat from the interior of the sun to the outer surface and what drives the weather on the surface of the earth.

Pop Quiz

When a pot boils on your stove, is heat being transferred by conduction or convection? Answer: Convection—you can see the places where the water bubbles up and sinks down if you watch the pot.

544 **Radiation.** If you hold out your hand to a fire, you feel heat. Your sensation is the result of an energy transfer from the fire to your hand. The heat is transformed into infrared radiation in the fire. This radiation travels through space and is eventually absorbed by your hand, where it is converted back into the kinetic energy of atoms. Every object whose temperature is above that of its surroundings loses heat by radiation.

The Second Law of Thermodynamics

545 **Energy is not the whole story.** To understand why, think about the fact that it is easy to scramble an egg and very difficult to unscramble it. Yet from the point of view of energy, one process is no different from the other. Our intuition tells us that it is hard to establish order in the world and easy to destroy it, but we clearly cannot relate this insight to the first law of thermodynamics alone.

546 **The second law of thermodynamics deals with the concept of order in the universe.** It is a difficult concept to understand, yet it rests on some of the simplest observations. If you put an ice cube on a table, heat moves from the air into the ice cube rather than vice versa. This, in fact, leads to one statement of the second law of thermodynamics— 1. Heat will not flow spontaneously from a cold to a hot object.

There are other ways of stating the law. They are:

2. It is impossible, even in principle, to convert heat into

work with 100 percent efficiency.

3. The disorder of an isolated system must increase with time or, at best, remain constant.

It is possible to prove mathematically that any one of these three statements of the second law implies the other two. In other words, although it certainly isn't immediately obvious, they are all equivalent statements.

547 **The world is going to hell in a hand basket,** at least according to the last statement of the second law. It tells us that the amount of disorder in a system must increase with time *unless* the system is not isolated from its surroundings. In technical language, this statement of the second law uses the concept of entropy. Entropy is a measure of the amount of disorder in a system at the atomic level, and the second law can be stated in the form that "the entropy of a closed system cannot decrease."

548 **The second law does NOT imply that systems cannot become more ordered.** If energy is flowing between parts of a system, it is possible for one part of the system to become more ordered while another part becomes more disordered. Take the earth and the sun as an example. A small number of molecules in living systems on the earth become ordered, but at the same time a much greater number of molecules on the sun become more disordered. The net disorder of the system increases.

549 **The second law says that we must think of energy in terms of its quality as well as its quantity.** If we cannot convert heat into work with 100 percent efficiency, then whenever we wish to make such a conversion (e.g., in an electrical generating plant), some of our original store of heat must be dumped into the atmosphere, where it becomes unrecoverable. Thus, we can think of high-grade energy as being energy associated with high temperature, and picture this energy as being degraded each time it is used.

An example might be the heat at the core of a nuclear reactor or a coal-fired power plant. This is converted into a lower-grade energy in the form of electricity which is brought to your home and then converted into still lower-grade energy when you use it to run an appliance. At each step, some of the energy is converted into low-grade heat and dumped into the environment. Although no energy is lost at any step in this process, the process is nevertheless a one-way street because, with each conversion, some energy is put into its lowest-grade form and cannot be used further.

550 **The second law has important consequences for energy policy** because it sets a maximum possible efficiency for any heat-driven engine. Such engines include those in electrical generating plants and those in cars. It turns out that, according to the second law, our current generating plants can be no more than about 40 percent efficient. In other words, when you burn a ton of coal to generate electricity, almost two-thirds of the energy in that coal must be dumped into the atmosphere. This is not a result of faulty design or poor efficiency on the part of the engineers—it is a consequence of one of the most fundamental laws of nature.

In point of fact, modern generating plants typically operate at 30 percent efficiency or better. The engineers have pushed the design very close to the theoretical limits.

551 **The use of solar energy to heat a home or the burning of oil for the same purpose does not run into second-law constaints, because an oil furnace is not an engine.** The purpose of an oil furnace is not to produce useful work, but to produce heat. Therefore, such devices deliver energy at much higher efficiencies than an electrical generating plant.

552 **The universe will not "run down" because of the second law,** despite the fact that during the nineteenth century people talked about something called the "heat death" of the universe. The idea was that because of the second law all available energy would eventually be dumped as waste heat and everything in the universe would approach the same temperature.

We no longer believe that this will be the fate of the universe. We now understand that the universe is expanding and cooling. Consequently, our view of the end of the universe is very different now than what it was in the nineteenth century.

Heat and Materials

553 **Most materials expand when heated** because when heat is added, the atoms move faster. You can think of this added motion as a requirement on the atoms' part that they have more "elbow room." As a consequence of higher temperatures, then, most materials expand.

554 **Water does not always expand when heated.** You are probably aware of the fact that when water freezes it expands. This is why water pipes burst in cold weather. What you may not be aware of is that this property characterizes not only the water-ice transition, but low-temperature water as well. From zero to four degrees centigrade, water contracts as it is heated. In other words, the water is at its most dense when it is at four degrees, and is actually less dense at lower temperatures. This means that water at the bottom of the oceans can be warmer than water higher up.

555 **Melting and evaporation both involve energy.** If you heat a solid, the atoms move faster and faster. Eventually a point is reached at which the atoms cannot be held in their rigid structure anymore and start breaking loose. When this happens, the material changes from a solid to a liquid. We say that it melts.

It requires extra energy to raise the temperature of a material past the melting point. This is because in addition to adding to the kinetic energy of the molecules, we have to break the bonds that were holding the atoms together. While the energy needed to break the bonds is being added to the systems, the material stays at a constant temperature (the melting temperature).

The same sort of thing happens if a liquid is boiled. In this case, energy needs to be added to allow molecules to escape from the attraction of their neighbors in the liquid and fly into the air.

556 **The boiling point of a liquid depends on the pressure.** The escape of molecules from the surface of a heated liquid is easier if the external air pressure is lower. This is why water boils at a lower temperature at high altitudes than it does at sea level. If you have ever tried to follow a recipe while living in the mountains, you have probably run into this phenomenon. A three-minute egg may have to be boiled for longer than three minutes in Denver.

557 **Absolute zero is the lowest possible temperature.** In the classical view, the lowest possible temperature would correspond to the stopping of all

The operation of this old-time still depends on the fact that the boiling point of alcohol is a little less than that of water.

atomic motion. Absolute zero is −2738 C (−4568 F).

Today, with the advent of quantum mechanics, we know that atoms cannot actually "stop" in the usual sense of the word. Instead, we define absolute zero as the temperature at which atoms have the lowest possible energy they can have consistent with the laws of quantum mechanics. A material at absolute zero in both the classical and quantum mechanical pictures is a material from which no further energy can be extracted by any means.

558 **The lowest temperatures produced in laboratories are less than a billionth of a degree of absolute zero,** and it is routinely possible to produce temperatures of 4 degrees above absolute zero. In fact, you can buy liquid helium at this temperature for about the price of cheap scotch. At the other end of the temperature scale, fusion reactions, either in weapons or in laboratories, take place at temperatures comparable to the interior of the sun, which is thought to be about 150 million degrees centigrade.

5

MODERN PHYSICAL SCIENCE

The Fermi National Accelerator Laboratory outside Chicago—modern science at its biggest. The ring is over a mile across.

The Electronic Properties of Matter

559 **How a material responds to electrical forces depends on how its atoms are arranged.** When atoms are locked together to form solids, liquids, or gases, the way the electrons are arranged depends on the fine details of the forces between the atoms. In particular, they determine whether the electrons are able to move around the material in response to external electrical forces, hence whether electrical current can move through the material.

560 **Conductors must contain free electrical charges.** A conductor is a material through which electrical current can flow. In order to be a conductor, a material must contain free electrical charges. There are many kinds of conductors, and they differ in the type of free charges available and in how they are created.

561 **In materials like metals, some electrons are not tied to their individual atoms,** but are free to move around throughout the material—in effect, they are shared by all the atoms. These loose particles are called "conduction" electrons. In a metal like copper, roughly one electron per atom is of this type. Metals are the most commonly used conductors.

562 **A material does not have to be a metal to** be able to conduct electricity. When you turn on a fluorescent light, some of the atoms in the gas become ionized and lose electrons. These free electrons can move when a voltage is applied. In the same way, if you dissolve salt in water, there will be charged ions floating in the water. These ions are free to move, and hence can constitute an electric current. Both salt water and ionized gases are examples of non-metallic conductors.

563 **In order to be widely used to carry electricity, a material must be a good electrical conductor and it must be relatively cheap.** Copper fulfills both of these requirements, which explains why it is the most commonly used conductor. The wires in your house are most likely made from copper. Aluminum is also used for this purpose occasionally, but it is not as good a conductor as copper. In situations where cost is no object (for example, in satellites in space), gold and silver are used in electrical circuits because they are slightly better conductors than copper, although they are much more expensive.

564 **In an insulator, electrons are tightly bound to their atoms.** An insulator is a material that won't carry electrical current. External electrical forces are not strong enough to pull the electrons away from their parent molecules in such a material. Consequently, no electrons move when a voltage is applied, and no current flows. Wood, plastic, rubber, and glass are all insulators.

Pop Quiz

What material is used to cover the switches in your house? Why is this material used, do you suppose? Answer: The most common material used in this way is plastic, which is an insulator. Insulators are used so that you cannot touch the circuit.

565 Every material is a conductor when the voltage is high enough, since a high enough voltage can always tear electrons loose from atoms. For example, air is normally considered to be an insulator. However, when a large electrical charge accumulates on a thunder cloud, the atoms of air in the region of the cloud are torn apart, turning the air into a plasma. This is how a lightning bolt is created.

566 There are a few materials—silicon and germanium are the two most common examples—which are neither pure conductors nor pure insulators. A typical piece of silicon might carry one-millionth of the electrical current that a similar piece of copper would carry. Materials like this are called semiconductors.

The secret of the semiconductor is that some of the electrons are bound very loosely to their parent atoms. When these atoms move around (as all atoms do at temperatures above absolute zero), some of these nominally bound electrons are shaken loose. Thus, there are a few conduction electrons in the material. This explains why semiconductors carry current, and also why they don't carry it very well.

567 Your personal computer, your TV set, your radio, and even the little dictating machine I am using at this moment all use devices made from semiconductors. You couldn't make a plane reservation or call your friends across the street without the use of semiconductor devices. Thus, this "oddball" type of material turns out to be absolutely crucial for the health of modern technology.

It's a good thing there are beaches full of silicon all around the world.

568 Both electrons and "holes" conduct electricity in semiconductors. When an electron is shaken loose, it carries current in the normal way. It also leaves behind a vacant space, called a "hole," in the lattice. An electron from other atoms can fall into this hole, creating a new hole in its parent atom, then another electron can fall into *that* hole and so on. The serial movement of electrons in one direction is equivalent to the movement of a hole in the opposite direction and this, in turn, is equivalent to the movement of a positive charge. The motion of the hole is a second kind of electrical current in a semiconductor.

569 **The technique which allows a semiconductor to be turned into a personal computer or other useful device is called "doping."** This is how it's done: when raw silicon is being melted, small amounts of other kinds of atoms are added to the melt. When the silicon solidifies, these atoms are incorporated into the crystal structure, with four outer electrons forming covalent bonds. If the impurity atom has more than four electrons in its outer shell, the extra electrtons will be free to move through the material. They will carry current, but will not leave holes behind on their parent atoms. The impurity atoms, having lost electrons, will have a net positive charge. This sort of system is called an "N" (for negative) type semiconductor, since the impurity donates a negative charge carrier.

If, on the other hand, the impurity has less than four electrons in its outer orbit, there will be a hole in the crystal structure, and this hole will be free to move without leaving an electron behind. When the hole in the impurity atom moves, the atom will have an extra electron and therefore a negative charge. This is called a "P" (for positive) type semiconductor.

570 **The simplest semiconductor device that can be built is a diode,** which is an N- and a P-type semiconductor brought up against each other. The donated electrons from the N-type semiconductor will diffuse across the boundary and fall into the holes in the P-type semiconductor, filling them up. Similarly, donated holes from the P-type semiconductor can diffuse the other way and absorb donated electrons. The net result is that all the donated particles and holes cancel each other out, leaving a line of positive ions on the N side and a line of negative ions on the P side.

By manipulating the atoms, we have created a region that can act as a battery on the atomic scale. An electron placed on one side of that boundary will be repelled by the negative ions and attracted by the positive. The electron will therefore be accelerated across the boundary, acquiring energy in the process, as if it were in a battery. The difference between this atomic battery and the battery in your car is that the atomic battery, once created, cannot run down. So long as the atoms stay in place, the atomic "battery" will continue to work. This simple manipulation of atoms is the technology that lies behind most of the microelectronics industry.

571 **The most ubiquitous semiconductor device is the transistor.** A typical transistor is just three slabs of semiconductor—either two P-type semiconductors sandwiching an N-type, or the reverse. Charges (either electrons or holes) flow from one piece of "bread" in the "sandwich" to the other piece. For a brief period, they are in the small semiconductor of a different type (the "meat"). When we make small changes to the voltage that is applied to the central part of the device, therefore, it can produce large changes in the current that is flowing between the two end pieces, much as small changes in a valve can produce large changes in the flow of water in a large pipe. In this manner, a small signal applied to the center part of the transistor "sandwich" can be translated into a large signal in the current flowing through the transistor. This is how a transistor in your stereo system takes a small signal (the electrical signal from the stylus) and converts it into a large signal (the current that drives the speakers). A device that takes a small signal and converts it to a large one is called an amplifier.

572 A microchip may contain thousands of transistors. Although the original transistors were fairly bulky devices, modern technologists can fit thousands of transistors onto a silicon wafer the size of a postage stamp. This is done by a process in which N- and P-type materials are deposited in layers directly onto the silicon. In this way, the layers of the "sandwich" can be made to be extremely thin—in principle, as little as one atom across. The resulting assembly of semiconducting devices, called a "microchip," is what operates inside your hand calculator and your personal computer.

573 The actual working part of your hand calculator—the microchip—is smaller than a postage stamp. The calculator can't be made much smaller than the credit-card-sized things now available, however, for one simple reason. The keys have to be big enough so that data can be punched in by a human being. Thus, the limiting size of the calculator is not set by semiconductor technology, but by the size of the human finger!

574 It is easy to forget how recently our technology has been revolutionized by the advent of semiconductors. When I was a senior physics major at the University of Illinois in the early 1960s, one of my favorite classes was a state-of-the-art electronics lab. As a special treat for students who finished the regular labs early, the instructor had arranged for us to get one of those newfangled devices called transistors and mess around. Since that time, electronic systems run with vacuum tubes (as they were in the "old days") have become museum pieces.

575 There are materials called superconductors that carry electrical current with no loss of energy. In superconductors, in other words, there is no heating, and electrical current, once started, will flow forever. Until very recently, such behavior could be seen only in the materials kept at a temperatures a few degrees above absolute zero. A standard commercial superconductor has to be immersed in liquid helium (at 4 degrees above absolute zero) to keep it from "going normal."

Here's how most superconductors work: When a single electron goes between two positive ions in a non-superconducting material, the ions are attracted toward that electron and move slowly toward each other. The two ions constitute a positive charge which can then attract another electron. In this way, two electrons form into a pair, one trailing the other. When the temperature is low enough, the normal thermal motion of the ions will not disrupt the pairing, and all the electrons in the material combine into these so-called Cooper pairs. The pairs interlock and intertwine so that, in effect, all of the electrons in the metal are bound together in a single structure. In this kind of situation, no electron can lose energy and slow down because it is locked into the other electrons. Thus, the entire complex of electrons move through the lattic without transferring any energy to it.

576 If all the electrical current used in the United States could be carried on a single superconducting wire, the wire would be no bigger across than a basketball.

577 The main use of superconductors up to this point has been as magnets. The benefit of using a superconducting wire in creating an elecromagnet is obvious: once you get a current started in a loop of wire, it will keep on flowing (and therefore keep on producing the magnetic field) forever, provided only that the wire is kept cold.

578 Room-temperature superconductors are a Holy Grail in physics. The great technological drawback to the use of superconductors, of course, is the fact that the phenomenon ocurs only at very low temperatures. For decades, physicists have tried to build materials that would be superconducting at normal temperatures. Although there are no immediate prospects for such a "room-temperature" superconductor, it remains an important dream.

In 1986, materials were discovered that remained superconducting at "high temperatures"—high, that is, compared to the temperature of liquid helium. Materials have been found that remain superconducting well above 100 degrees kelvin (-170 degrees celsius). Such materials would remain superconducting in a bath of liquid nitrogen, a material that is much cheaper and more plentiful than liquid helium. These materials tend to be brittle ceramics, however, and it will probably be a while before they are produced in commercially useful form.

The Nucleus of the Atom and Radioactivity

579 **Most of the mass of the atom, but almost none of its volume, lies in the nucleus.** In a typical atom, the nucleus will weigh about four thousand times as much as the electrons. Therefore, to a good approximation, we can ignore the electrons when we talk about the mass of the atom.

On the other hand, the atom is almost all empty space. If the nucleus of an atom were a basketball on the floor in front of you, the electrons would be like a few dozen grains of sand scattered around the county in which you live. For experts, we point out that the linear dimension of the nucleus is typically 10^{-5} times the linear dimensions of the entire atom.

580 **The nucleus was discovered in 1911** by Ernest Rutherford in Manchester, England. He and his coworkers took radiation known as alpha particles (see the following) and allowed them to hit a thin gold foil. Although most particles went through or were only slightly deflected, one particle in a thousand was bounced backward from the atoms in the foil. Rutherford compared the experiment to the process of shooting a bullet into a cloud of steam and occasionally finding a bullet bouncing back. The only conclusion that can be drawn in either case is that some-where inside the atom (or cloud of steam) was a small dense body capable of deflecting fast-moving particles and making them change direction. Rutherford called this small, dense body the nucleus.

581 **Rutherford is one of those unusual individuals who made his most important contribution to science *after* he received the Nobel Prize.** He got the prize in chemistry in 1908 for working out the nature of the particles given off by radioactive materials (see the following), then went on to discover the nucleus.

582 **The nucleus is made from protons and neutrons.** Rutherford called the nucleus of hydrogen—a single particle with a positive electric charge—the proton ("the first one"). The total positive charge of the nucleus, then, is the sum of the charges on the protons, and the number of electrons in orbit in a neutral atom is equal to the number of protons in the nucleus.

The neutron ("neutral one") is about as massive as the proton but, as the name suggests, has no electrical charge. It adds to the mass, but not the charge, of the nucleus.

Most stable nuclei have approximately equal numbers of protons and neutrons. When this general rule is broken, as it is for heavy elements, the tendency is for nuclei to have more neutrons than protons.

583 The chemical identity of an atom depends on the number of protons in the nucleus. The number of protons in the nucleus (the so-called atomic number, typically denoted by the letter Z) determines the chemical nature of the atom, because this chemical nature is determined by the outermost electrons in the atom. Thus, if you tell me the number of protons in a nucleus, I will tell you what kind of atom you are talking about. For example, if there are six protons, the atom is carbon; if there are eight, it is oxygen, and so on.

584 Extra neutrons do not change the chemical nature of an atom because they do not change the number of electrons needed to cancel the charge of the nucleus. It is therefore possible to have many different species of a given kind of atom, each having at its center a nucleus with the same number of protons, but a different number of neutrons. Two atoms whose nuclei have the same number of protons but a different number of neutrons are said to be isotopes of each other.

Pop Quiz

A*n atom of deuterium has a nucleus with one proton and one neutron. Of* what chemical element is deuterium an isotope? Answer: Hydrogen.

585 To a very good approximation, you can regard the electrons and nuclei as two separate systems, each doing its own thing and ignoring the other. This means that it makes very little difference to the nucleus whether the atom is by itself in space or whether its outer electrons are part of chemical bonds. The nucleus will do whatever it's going to do in both situations.

It also means that it makes very little difference to the electrons whether there are extra neutrons in the nucleus or not. Different isotopes of a given element are equally adept at finding places in minerals and other materials, and all the isotopes of a given element will appear in any material that incorporates that element.

Radioactivity

---

586 A nucleus is radioactive if it emits particles spontaneously. Familiar nuclei are stable—that is, they will not spontaneously change from one form to another. There are, however, nuclei which are not stable. Uranium is probably the most familiar example of such a nucleus. These nuclei spontaneously emit particles which we call "radiation." A nucleus that emits radiation is said to be "radioactive" and the act of emitting radiation is said to be "radioactive decay."

Madame Curie, one of the founders of nuclear science.

587 Marie Sklodowska Curie, a Polish woman who spent most of her professional life in France, played a major role in early studies of radioactivity. There are many remarkable things about her: she is the only person who has ever won two Nobel prizes in scientific fields, she is the discoverer of the elements radium and polonium, and she is one of the founders of the study of radioactivity and, hence, of nuclear physics.

So great was the resistance to the idea of a woman scientist in the late nineteenth century, however, that despite her two Nobel prizes, *she was never elected to the French Academy of Sciences!* You'd think that after the first one, the boys would have gotten the message.

588 There are three kinds of radiation. The physicists at the time had no idea what these radioactive particles were, so they gave them names to express their mysterious nature—they called them alpha, beta, and gamma rays, respectively.

Alpha particles are made of two protons and two neutrons—they are actually the nuclei of ordinary helium. Beta radiation is made up of electrons. Because this mysterious new radiation was discovered only shortly after the electron itself, the fact that beta rays and electrons are identical wasn't appreciated for a number of years. The decay of a nucleus which results in the emission of an electron is called "beta decay." Gamma radiation is ordinary X rays emitted when the

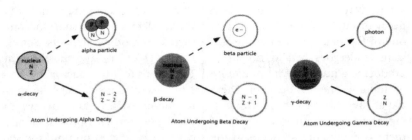

Alpha (left), beta (center), and gamma (right) radiation.

protons and neutrons rearrange themselves inside the nucleus.

589 **The helium that you use to inflate your children's birthday balloons (and which is used in liquid form to keep superconductors cold) is not taken from the earth's atmosphere.** Instead it comes from radioactive decay of nuclei deep within the earth. These decays produce alpha particles that slow down, acquire electrons, and form helium that is then trapped along with oil and natural gas. When oil and natural gas reserves are developed, the helium is separated and sold.

590 **The neutron itself undergoes beta decay.** In fact, if you were to watch a free neutron, you would see it "fall apart" in about eight minutes. The end products of the decay are a proton, a neutrino, and an electron. For technical reasons, a neutron that is safely ensconced in a nucleus can be stable and safe from beta decay so long as it stays there.

591 **The energy involved in radioactivity comes** **from the conversion of mass.** If you measured the masses of the final products of a radioactive decay, you would find that they had less mass than the original nucleus. The difference in the before-and-after masses is converted to energy according to the equation $E = mc^2$, and it is this energy that we see as the energy associated with radiation.

592 **Radioactive decay is the ultimate philosopher's stone**—the stone that alchemists believed could turn lead into gold or, more generally, one chemical element into another. Since alpha and beta decay change the number of protons in a nucleus, they also change the chemical identity of the atom of which that nucleus is a part.

After alpha decay, a nucleus will be able to hold two fewer electrons than it did before the decay. The two "extra" electrons will eventually wander off, leaving behind an atom that has two fewer electrons in orbit. This atom will, of course, be identified as a member of a different chemical species from the original atom.

One way of thinking about beta decay of a nucleus is to imagine that one of the neutrons inside the nucleus undergoes beta decay itself, producing a nucleus with one more proton and one less neutron. There are always loose electrons wandering around in nature, and one of them is eventually attracted to the atom. The final result is that a new chemical element has been born— one with one more electron in orbit than had originally been there. Again, a new chemical element has replaced the old one.

Since gamma decay just involves a reshuffling of protons and neutrons, it does not change one element into another

593 **Uranium 238 decays by alpha emission.** Uranium has 92 protons in the nucleus, so the daughter nucleus of this decay will have 90 protons and a total mass of 234 (238−4). The product will, in fact, be what chemists call Thorium 234 (^{234}Th).

594 **The decay of a single nucleus is seldom the whole story in a radioactive decay,** because more often than not the daughter nuclei—the results of the decay—are themselves radioactive. Thus, the original decay gives rise to a daughter which decays, and *that* nucleus decays into still another and so on. This string of events is called a decay chain. The chain keeps going until a stable nucleus is produced.

As a consequence of the existence of decay chains, a pure sample of a given element will soon become mixed with other elements. For example, U-238 decays into thorium. The thorium then decays by beta emission into an element called protactinium, which also decays by beta emission. This process of successive decay goes on until the stable nucleus lead 208 is reached.

595 **Americans are becoming aware of the health risks associated with the accumulation of radon in homes.** Radon is one of the elements in the decay chain that leads from uranium to lead. Thus, it is always being produced by nuclear decays in the ground. Once an atom of radon is produced, its future movement is governed by its chemistry which, in this case, dictates that it does not bind to materials around it but instead seeps up through the ground and into basements of houses.

596 **Radioactive nuclei do not decay all at once.** They decay at what appear to be random intervals. Watching these nuclei decay is a lot like watching popcorn pop in a griddle, with the kernels exploding at different times.

The number that is usually used to measure the speed with which radioactive nuclei decay is called a half life. It is defined to be the time that it takes for half the nuclei of a given material to undergo decay. Thus, if you start with one thousand atoms, the half life is the time you have to wait until you have only five hundred left.

Half lives of nuclear isotopes can range from billions of years to microseconds. Some examples:

uranium 238	4.6 billion years
carbon 14	5730 years
radon 222	3.8 days
uranium 239	23.5 minutes
carbon 10	19.4 sec

Enduring Mystery

597 **We do not know how to predict the half life of radioactive nuclei.** Although half lives of radioactive nuclei can be (and are) measured quite accurately, we do not yet have the computing power to be able to predict what the half life will be of most nuclei. This problem is simply too complex to be handled by even the largest computers now available to us.

598 **Radioactive materials are "hot" in two senses of the word.** Radioactive materials are "hot" in the sense of giving off radiation, but they're "hot" in the normal thermal sense as well. You can see why by asking what happens to something like an alpha particle after it is emitted in a radioactive decay. The alpha particle will move out into the surrounding material and bounce around, something like a ball bearing in a pinball machine. As a result of these collisions, the initial energy of the alpha particle will be shared among the atoms in the surrounding material, which will move faster as a result. This faster motion, of course, is what we perceive as heat. Thus, any material in which radioactive elements are present will be heated up by the presence of these elements.

Heat generated by radioactivity is thought to contribute significantly to the operation of plate tectonics. The generation of heat by radioactive materials is also important in the problem of disposing of radioactive wastes, because the wastes must be stored in materials that will not melt over long periods of time.

599 **If you took a cube of granite that you could hold in your hand and kept the heat generated by radioactive decay of nuclei normally found in the rock from escaping,** within a million years enough heat would be generated to melt the rock completely.

600 **Radioactivity is not "unnatural."** Because human awareness of radioactivity is recent, and because the public has really become conscious of radioactivity only since the Second World War, there is a tendency for many people to believe that radiation is something new in the human environment. In fact, our race lives and evolved in an environment that is full of radioactivity. Uranium, which undergoes radioactive decay, is a common element in the earth's crust. It is more common than things like silver and mercury. The decay chain which it initiates (see the foregoing) fills the earth with radioactive nuclei. In addition, the earth is constantly being bombarded by cosmic rays. These particles, mostly protons, are generated in the sun and other stars and rain down on our atmosphere all the time. There, in collisions with molecules in the air, they produce showers of particles. At this very moment, particles from these showers are passing through your body at the rate of about three a minute and adding to the natural background radiation levels.

Thus, when you read about radioactivity being discovered somewhere, you should be aware that most places already have radioactive materials in them, and the question you should ask is not "Is it radioactive?" but "Is it more radioactive than it would normally be?"

601 All nuclear energy comes from conversion of mass.
Whenever a nucleus changes state, there will be a small change in its mass. If, as is the case for decays (see the foregoing), the sum of the masses present at the end of the change is less than it was before, this difference will be converted into energy.

There are two kinds of processes (other than decay) that are usually associated with the term "nuclear energy." These are fission and fusion.

602 Fission is a process that splits a large nucleus into two or more smaller daughter nuclei.
Fission of some nuclei releases energy, and fission of others requires an energy input. The most well-known energy-producing fission is that of uranium 235, which, when struck by a slowly moving neutron, splits and produces energy (in the form of the motion of high-speed fragments) and a few more neutrons, each of which can go on to produce more fissions. The result: a continuing release of energy so long as the supply of uranium 235 lasts.

603 There is a good chance that the electricity in your home was generated in a nuclear reactor.
This

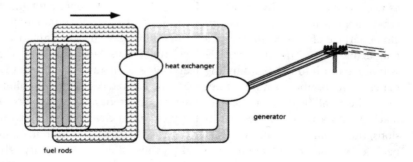

A fission reactor.

is how a reactor works (see sketch): uranium 235 is contained in fuel rods about as big across as a pencil. Neutrons produced in fission reactions leave their "home" fuel rod, are slowed down by the water or fluid in between the rods, and initiate a fission reaction in another rod. The effect of all these reactions is to heat the water, which is then piped outside the reactor proper and used to heat water in a separate system of pipes. Steam from this secondary system drives the generators that produce electricity.

604 **Fusion is the process by which two small nuclei come together to form a single large nucleus.** The most important fusion reaction is that which produces helium from hydrogen. This is what powers the sun and other stars. It is also this reaction that we have been trying to harness in the laboratory so that we can use it as a source of electrical power.

605 **Historically, attempts to harness fusion have involved trying to reproduce the conditions in which fusion takes place in the interior of the stars.** This involves heating a hydrogen gas to very high temperatures and compressing it in magnetic fields until the fusion starts. This process is called "hot" fusion. We have not yet been successful in producing a self-sustaining, controlled fusion reaction.

In 1989, scientists claimed to have discovered the possibility of another route to fusion—the so-called "cold" fusion process. The failure of other scientists to reproduce the original results has led to an abandonment of this road to fusion.

Quantum Optics

606 **Modern optics is quantum optics.** Through the 1950s, optics was considered by physicists to be a rather moribund field of study—there seemed little chance that interesting new knowledge would come from it. Little by little, it was being dropped from the university curriculum.

This situation has changed. With the introduction of the laser (see the following), the field has boomed. Modern optical equipment depends primarily on the quantum properties of light, and is therefore usually referred to as "quantum optics" to distinguish it from classical optics, which treats light purely as a wave.

607 **"Laser" stands for Light Amplification by Stimulated Emission of Radiation.** Here's how it works: we start with a system in which many atoms have electrons in excited states. If there is a photon whose energy is exactly the same as the energy difference between the allowed orbits in the atoms, the presence of that photon stimulates electrons in upper orbits to make a downward transition and emit another photon. The emitted photon will have the same energy as the first, and they will be aligned crest to crest. The second photon, in turn, stimulates emissions from atoms, as do subsequent photons in turn. The result: a single photon will produce a flood of identical photons on the material.

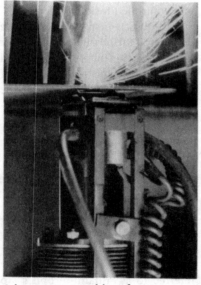

A laser cuts metal in a factory.

At each end of the collection of atoms is a set of parallel mirrors. Photons encounter those mirrors and are reflected back through the material. Thus, each photon may bounce back and forth billions of times, amplifying itself as it goes. The important point is that only those photons moving exactly perpendicular to the two mirrors will remain within the laser. Any photon moving at an angle will make a couple of reflections and then walk itself out of the sides.

The net result of this process is that you have floods of precisely aligned photons bouncing back and forth within the laser. At each bounce, some photons leak out, and it is this "leakage" that we perceive to be the beam of the laser.

608 **Lasers need to be "pumped."** After photons are emitted, as described above, energy needs to be added and the electrons have to be moved back to the excited states.

This can be done in any number of ways. The laser can be heated so that the atoms get new energy through collisions. Light can be introduced into the laser from another source—in some lasers, this is done with equivalent of a camera flash bulb. It is even quite common these days to use one laser to "pump" another. But whatever process is used, a laser needs a continuous input of energy to operate.

609 **Lasers have many uses.** In the 1960s, lasers were an exotic tool developed by laboratory scientists to help them carry out basic research on the structure of atoms. Today, they are used everywhere in our society—in medicine (as surgical tools, for example), in surveying, and in manufacturing, to name a few examples. There are even trivial uses—I am often handed a laser to use to point to slides when I give lectures. This is as good an example as you can have of how seemingly useless basic research can have a profound impact on the economy.

610 **The world's largest laser (maybe) is the ALPHA,** run by TRW Corporation. It achieves a continuous output of several hundred kilowatts, so it's certainly one of the most powerful in the world. Since most high power lasers are connected with defense work, however, information about them tends to be classified. There are surely devices more powerful than the ALPHA, but no one's talking about them.

611 **Fiber optics systems can move light around corners.** When light moves from a more dense to a less dense medium (e.g., when it moves from glass into air), the phenomenon of refraction takes a strange turn. If light approaches the interface between the two media at a grazing angle, it never goes into the air but is reflected back into the medium. This phenomenon is called "total internal reflection."

If light is introduced into a cylinder of glass at just the right angle, it can never get out because each time it approaches the surface it is reflected back. The light is trapped and can only rattle its way down the cylinder. Even if the cylinder is a fiber wrapped around corners and tied into knots, the light can't leave. It is this aspect of the behavior of light that gives rise to the new technology of fiber optics.

612 **Images can be formed with fiber optics.** The picture on a TV set is made up of millions of black-and-white (or colored) dots that our eye integrates to produce a smooth image. Fiber optics works the same way. A large number of small fibers are wrapped into a bundle. Light from an object falls on the exposed ends of a bundle, and the light that enters each fiber is transmitted unchanged to the other end. Thus, at the far end of the fiber optics you get a series of dots which, like the picture on the television set, is integrated into a single image by your eye.

613 **You are likely to run into fiber optics in your hospital.** It used to be that

A fiber optic cable bends light in loops.

surgeons would have to cut into a joint to see what was going on inside. Now they can make a small incision and insert a tube the size of a pencil into it. The tube contains a light and a fiber optic bundle. The light from the bundle is brought out as described in the previous item and made into a TV picture. In this way, the surgeon can see the inside of a knee without having to do invasive surgery. In fact, fiber optics is often combined with miniaturized tools that can be inserted through the same tube as the fiber optic cables and manipulated from outside the body.

614 When you read about a running back suffering torn knee cartilage and undergoing "arthroscopic surgery," you are reading about an application of fiber optics and miniaturized tools. The knee is repaired without being opened up, and the patient can be walking (albeit with a limp) the next day and often can return to full activity within a matter of weeks. Thus does science enhance the operation of the National Football League.

615 Fiber optics is used in communication. It used to be that when you made a telephone call, your voice was converted to electrical signals which were sent along ordinary copper telephone wires. Such wires carry a relatively small number of messages at a time. Now, the sound of your voice can be converted to signals carried by light in a fiber optic system. Because the wavelength of this light is very short compared to the wavelength of ordinary electrical signals, many more messages can be crammed into a single cable than could be carried by conventional means. In the United States today, much of the long-distance traffic between major cities, particularly on the East Coast, is carried through fiber optic cables rather than copper wire or via satellite links.

616 In 1989, the first transAtlantic optical cable was laid under the ocean. Now, your phone calls to Europe may well be traveling along those very fibers.

Elementary Particles

617 There are hundreds of elementary particles. Although we usually picture the nucleus as being a static collection of protons and neutrons, in fact, it is a dynamic place. All kinds of particles are whizzing around, bashing into each other, and being created and destroyed as their energy is converted into mass and their mass is converted into energy. Being inside a nucleus is probably more like being in the middle of a fireworks display on the Fourth of July than anything else.

Discovering and sorting out all of the particles that live a fleeting life inside the nucleus has been the major task of elementary particle physics, and in the period since the 1950s, over two hundred of these particles have been discovered.

618 Learning about "elementary" particles can be compared to peeling an onion. In the last two centuries, we've gone through one layer of "elementary" structure after another, always to find that there's another structure underneath. First we got down to the atoms. Then we split the atoms and found that there were nuclei. Then we took the nuclei apart and found that there were elementary particles. Today we think these "elementary" particles

are made of quarks. The question, of course, is whether we've got to the end of the onion, or whether there are lots more layers to be found.

619 Elementary Particle Physics is practiced with accelerators, or "atom smashers." These are machines that take either protons or electrons and push them until they are traveling at almost the speed of light, at which point they are allowed to hit a target. The idea is that you learn what's inside a nucleus by hitting it so hard that you knock it apart. The process has been compared to learning how to build a watch by dropping watches from the top of the Empire State Building and seeing what happens.

620 The SSC (Superconducting Supercollider) will be the world's biggest accelerator if it is built. Today,

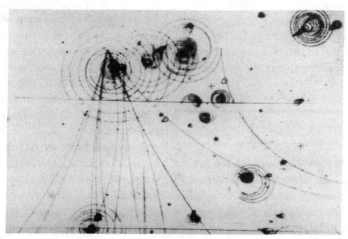

Interactions between elementary particles as seen in an accelerator laboratory.

the world's biggest particle accelerators are the Fermi National Accelerator Laboratory, located just outside Chicago, and the machine at the European Center for Nuclear Research (CERN) in Geneva, Switzerland. There are plans in the works to build a monstrous machine, over fifty miles in diameter, just south of Dallas. The project is given the acronym SSC, for superconducting supercollider. If it's built, it will cost in the neighborhood of $8 billion and will be so big that a good-sized city could fit inside its circumference. The purpose of the machine would be to explore the unification of the fundamental forces (see the following).

621 **There are two basic kinds of elementary particles**—those that live inside the nucleus, and those that don't. Particles that live inside the nucleus, like the proton and neutron, are called hadrons (from the Greek root for "strong"). Particles that live outside the nucleus are called leptons (from the Greek root for "weak"). Electrons in orbits far away from the nuclear maelstrom are examples of leptons.

Enduring Mystery

---

622 **Why are there extra leptons?** In addition to the electron, there are two other particles just like the electron except they are heavier. They are called the mu and tau mesons. Why nature should make the electron, and then go through the trouble of making it again and again, remains one of the enduring fundamental questions in physics.

623 **There are three kinds of leptons called neutrinos.** These particles have no mass, no electrical charge, and travel at the speed of light. One of my students once asked me, "If they don't weigh anything and they don't do anything, how do you know they're there?" Good question. In fact, the neutrinos were postulated to exist in the 1930s, but their existence wasn't verified until the 1950s, when their (very rare) interactions with ordinary matter were detected.

624 **Hadrons are made from quarks.** In 1964 two scientists at Caltech—Murray Gellmann and Fred Zweig—suggested independently that the "elementary" particles weren't really elementary after all, but were made from things that were more elementary still. At the time, they thought there would be three of these more elementary things, and they called them "quarks." The name comes from a line in James Joyce's *Finnegan's Wake*, "three quarks for Muster mark."

Since that time, we have come to know that there are six quarks instead of three. The names are fanciful—Up, Down, Strange, Charmed, Bottom, and Top (unless you come from Harvard, in which case the last two are Beauty and Truth).

625 No one has ever seen a quark, even though they have looked everywhere. Back in the sixties, this was taken to be evidence against the quark picture. Today, with their customary aplomb, theorists have turned necessity into a virtue and produced theories in which quarks are never supposed to be seen at all. The idea is that quarks are locked into the particles, and if you try to pull them apart, all you do is make more particles. It's almost as if the quarks were like the ends of rubber bands—if you stretch the rubber band and break it, you get two short rubber bands, but never the end of a rubber band all by itself.

626 Everything is made from quarks and leptons. First you put the quarks together to make elementary particles. Then you put the elementary particles together to make the nuclei. Then you tack the electrons on in their orbits to complete the atoms. Finally, you lock the atoms together to make all of the substances in the universe.

Most physicists think that with quarks and leptons we've discovered the basic structure of matter. A few rebels, on the other hand, think that quarks aren't "elementary" either, but are made of other things they call "preons."

While the history of elementary particle phsyics certainly doesn't predispose one toward confidence in final answers, I wouldn't hold my breath waiting for people to accept the idea of preons.

Enduring Mystery

627 Why are there as many quarks as there are leptons? There are six quarks and six leptons (the electron, the mu and tau mesons, and the three neutrinos). In fact, most modern theories incorporate this as one of the fundamental "givens" of the universe. But as to why there should be six of each, and why there should be as many of one as the other, no one has any idea.

628 Many terms are used to refer to the 200-plus elementary particles. Some of them are:

baryon—a particle whose decay products include a proton. The neutron, which decays into a proton, electron, and neutrino, is an example of a baryon.

meson—historically, a particle intermediate in mass between the electron and proton. Today, the term is used to refer to any particle whose decay products do not include a proton.

fermion—a particle whose spin is half integer (1/2, 3/2, 5/2 . . .) times a fundamental unit. Protons, electrons, and quarks are all fermions. Fermions are the building blocks of matter.

boson—a particle whose spin is an integer (0 or 1 or 2 . . .) times a fundamental unit. The photon is a boson. Bosons are involved in generating the forces that hold matter together.

629 There are four fundamental forces. In order of strength they are:

strong—the force that holds particles together in the nucleus and that, at a deeper level, holds quarks together in a particle.

electromagnetic—the force that acts between charges and magnets.

weak—the force responsible for some radioactive decay processes. One example is the beta decay of the neutron.

gravity—the attractive force that one piece of matter exerts on another.

630 Although the four forces appear to be very different, they are regarded by physicists as simply different aspects of a single fundamental force. When forces (like electricity and magnetism) are seen to be a single force, we say that the forces become unified, and theories that exhibit this fundamental unity are called "unified field theories." The accelerator at CERN (see the foregoing) has already demonstrated that the weak and electromagnetic forces become unified at energies we can reach in our laboratories. At the energy available in that machine, the differences between the electromagnetic and weak forces disappear. We expect that at much higher energies the strong force will unify with these two, and that finally all four forces will become unified.

631 The current best theory of elementary particles is called the Standard Model. Developed in the 1970s, this theory describes the unification of the weak and electromagnetic forces in a world where there are six quarks and six leptons. The unified weak and electromagnetic force is called the "electroweak" force. The Standard Model has been well verified by experiments at accelerator laboratories.

632 The Grand Unified Theory (GUT) and the Theory of Everything (TOE) are the frontier of modern elementary particle physics. GUT is the name given to the theories that describe the unification of the strong and electroweak forces. The various versions of the GUT have had some experimental success, but have failed in one important way. They predict that the proton, hitherto thought to be stable, will decay with a half life many orders of magnitude longer than the lifetime of the universe. This decay has not been seen. Thus, the GUTs are still in the process of being tested.

Tongue firmly in cheek, particle physicists have attempted to produce a theory in which all four forces are united. In this Theory of Everything (or TOE), there is only one fundamental force acting between particles. The world becomes as simple as it can be—one kind of force and one kind of particle.

Chaos

633 **A chaotic system is one in which the final outcome depends very sensitively on the initial conditions.** White water in a stream is a good example of a chaotic system. If you start a chip of wood at one position, it will come out at a particular point on the other side of a rapids. If you start the second chip of wood at a position almost (but not quite) identical to that of the first, the second chip will—in general—come out of the rapids far from where the first one did. The final outcome (the position of the chips) thus depends sensitively on the initial conditions (the place where they started their journey).

634 **For all practical purposes, the behavior of chaotic systems cannot be predicted.** It is impossible to measure the initial conditions of a system with perfect accuracy. The position of a wood chip at the beginning of its journey, for example, can only be determined as accurately as that best ruler available can measure. Since the final position of the wood chip will be very different if the chip is moved by an amount smaller than even this small margin of error, it follows that there is no way to predict where a wood chip will wind up.

Physicists and writers often express this point by saying that chaotic systems are "unpredictable."

They do not mean that if we know the state of a system exactly we cannot predict where it will be at some time in the future—these sorts of predictions are made all the time by computer modelers. What they mean is that because it is never possible to make a perfect set of measurements to determine the initial state of a chaotic system, their future states can never be predicted.

635 **The first discovery of a chaotic system was made by Edward Lorenz,** a meteorologist at MIT, who was forced to interrupt a long computer calculation of weather patterns. Rather than start the calculation again from scratch, he stored some intermediate results from the original computer run, then fed them in to let the computer pick up where it had left off. Much to his surprise, the results he got this way were very different from the results that he had gotten previously by running the calculation all the way through.

He discovered that the difference between the two sets of calculations came because the computer rounded off the numbers slightly differently when it stored them than when it was continuing to use them in a calculation. The rounding error in the computer produced a difference in the *eighth* decimal place in the relevant numbers. This was our first inkling that important systems in nature, such as the atmosphere, can be extremely sensitive to small changes.

636 Computers are the primary tool for studying chaos, and much of our understanding of chaotic systems comes from running computer models that trace these systems through time. A typical research problem would be something like this: an equation that describes a system is written down and solved on the computer. The starting point of the calculation is then changed slightly and the calculation repeated. If the predictions in the two solutions are wildly different, the system is chaotic and more detailed investigations are done.

637 Chaotic systems are nonlinear. Chaos is different from the kind of physics we're used to because the equations that describe chaotic systems are nonlinear. In a linear equation—the kind that describes familiar physics—one thing changes in direct proportion to another. For example, when you turn up the volume on your stereo system, twice as much turn gives you twice as much volume. In a nonlinear system this simple kind of relation does not hold. It's similar to what you get in your stereo system when you turn it up too loud and suddenly you get whistles, distortion, and all kinds of strange things. For technical reasons, the solution of nonlinear equations is a very difficult business, largely impossible to achieve without computers.

638 Fractals comprise another phenomenon that arises in nonlinear systems. The word "fractal" is a contraction of "fractional dimension." Consider the simple example of a coiled garden hose. From far away, it has zero dimensions—it's just a point. Closer up, it is seen as a solid object and therefore has three dimensions. Finally, from inside the coil, the hose becomes one dimensional, since we can specify any location on it by saying how far it is from the end. Thus, depending on our point of view, the dimensionality of the hose goes from zero to three to one dimensions. Fractals are a way of dealing with what happens in between.

Fractals can arise in nonlinear systems. An example of a fractal is given as follows. Start with a triangle and then in the middle of each side of the triangle draw another triangle. Then keep doing this on every straight line forever. It is obvious that if you look at any piece of this system at any level of magnification, you will see the same things—that is, a straight line with triangles on it. It is also obvious that there is a connection between the appearance of things at different scales of magnification. In fact, if you think about it, you'll realize

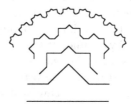

The generation of a fractal.

that you could not tell, just by looking at a line, what the magnification actually was.

639 **How long is the coast of England?** With this question, Benoit Mandelbroit introduced fractals to the physics community. He argued that if you looked at the coast of England from the point of view of a mapmaker, you would get a certain length. If you looked at it from an airplane, you would see little coves and inlets that might not have been obvious on a big map. If you got down and walked, you would see irregularities that would not have been visible from the airplane. If you started looking at the shoreline with a microscope, still more irregularities would be seen, right down to the level of individual atoms. Nevertheless, it's not hard to imagine that the coastlines produced by each of these operations could look alike. Thus, the coast of England is an example of fractal geometry.

A Personal Opinion

Chaos has been vastly oversold. Thanks largely to James Gleick's excellent book *Chaos*, the public has become aware of chaos and of its importance in science. Unfortunately, I think the whole field has been oversold. Some people have the impression that it is a major revolution in our thinking and will completely change the way we deal with the physical world. However, nothing could be farther from the truth. Chaos is very likely to provide insights into such problems as turbulence and growth of living systems. But it is unlikely to have much to say about the vast majority of physical phenomena, for the simple reason that these phenomena, already well studied, are known to be pleasantly linear and predictable.

The Atom

640 **The atom is the smallest unit of matter that retains its identity as a chemical element.** The name comes from the Greek for "indivisible," and we retain this name even though we now know that the atom is composed of still smaller particles.

The modern atomic theory dates to the English natural philosopher John Dalton (1766–1844), who published a book called *New System of Chemical Philosophy* in 1808. In this book, Dalton proposed something very much like our modern theory. He argued that for each chemical element there is a different kind of atom, and that different materials (what we would today call chemical compounds) are simply different combinations of these atoms.

641 The first modern model of the atom was proposed by Niels Bohr, a young Danish physicist, in 1912. The model is now known to physicists as the Bohr atom. The central feature of the Bohr atom is that electrons can be in orbits at only certain well-specified distances from the nucleus. The orbits at these distances are called "allowed orbits" or "Bohr orbits."

It requires energy for electrons to move from a lower orbit to a higher one, since work must be done to overcome the attractive force exerted on the electron by the nucleus. Thus, energy must be added to the atom to move the electron in this direction. Conversely, if an electron moves from a higher orbit to a lower orbit, there is an excess of energy which must be disposed of.

Note that different atoms have different Bohr orbits, since the energy of an electron depends on the forces exerted on it by the nucleus and by the other electrons, and both of these vary from one element to the next.

642 Emission of light corresponds to a transition from a higher to a lower Bohr orbit. If for some reason an electron finds itself in a higher orbit, it can spontaneously jump down to a lower one. When it does so, the difference in energy between the initial and final orbits leaves the atom in the form of a photon. This is the process by which an atom emits light and other forms of electromagnetic radiation.

643 When an atom absorbs light, electrons are moved from lower to higher Bohr orbits. The energy of a photon can be absorbed by the atom and used to move the electron from a lower to a higher orbit.

644 The existence of Bohr orbits explains the fact that different atoms give off different colors of light. When an electron moves from one orbit to another, it must absorb or emit only a specified amount of energy. This, in turn, means that any atom is capable of emitting and absorbing only those same discrete amounts of energy. Since the energy of a photon is related to the wavelength, and hence the color, of its light, each atom can emit and absorb only certain colors. This is why neon lights are red and sodium vapor streetlamps are yellow.

645 What an atom emits it also absorbs. Absorption of light by an atom corresponds to moving an electron upward between two orbits, while the emission of that wavelength of light corresponds to moving an electron downward between the same two orbits. Since the energy difference between the orbits does not depend on the direction of the quantum leap, it follows that if an atom can emit a certain color, it must also be able to absorb it.

646 The colors emitted by an atom amount to an "atomic fingerprint," because

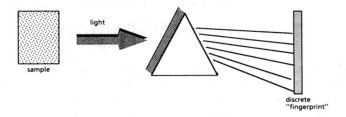

The fingerprint of an atom is carried in the light it emits.

no two elements have exactly the same Bohr orbits. This fact is the basis of the branch of science known as spectroscopy.

The fact that each atom emits and absorbs a different set of colors allows us to identify the presence of that atom in small samples of material. A sketch of the type of instrument that could be used to analyze light from a sample is shown. Different colors of light emitted by a sample are spread out by a prism to give us the "fingerprint" of that sample on a photographic plate or (more usually) an electronic detector. The instrument is called a spectroscope. This fingerprint is different for every atom and molecule.

647 **Spectroscopy helps astronomers.** In the early nineteenth century, Auguste Comte, the founder of modern sociology, published a list of things which he said it would always be impossible to do. High on that list was the analysis of the chemical composition of stars. In fact, the development of spectroscopy in the nineteenth century allowed us to do just that. By looking at light emitted from stars, we can detect the atomic fingerprints of atoms present, even though the star may be millions of light years away and we can never get our hands on a piece of it.

648 **Modern physicists have a strange picture of the atom.** Because quantum particles like the electron are thought of in terms of wave functions rather than classical particles, they think of the electrons as fuzzy clouds surrounding the nucleus, rather than as things analogous to planets circling the sun. The places where the clouds are most dense are where the electron is most likely to be found.

Quantum Mechanics

649 In the world of the atom and its components, everything comes in bundles. *Quantum* is Latin for "so much" or "bundle." In the interior of atoms, everything—mass, electrical charge, energy, momentum, and so on—comes in bundles. Nothing in this world is smooth and continuous.

"Mechanics" is the old-fashioned term for the science of motion, so "quantum mechanics" is the branch of science devoted to describing the motion of things in the subatomic world.

650 The biggest problem people have in dealing with quantum mechanics comes from our unconscious assumption that things will behave in the same way in the quantum world as they do in the ordinary world of our experience. Our intuition about how things ought to behave is based on our experience with large objects moving at normal speeds. There is no reason to expect that when we look at very small objects or objects moving at very high speeds, they should behave in the same way as do the objects with which we're familiar.

651 In the quantum world, you cannot observe something without affecting it. In Newtonian mechanics, we assume that we can observe something like a billiard ball or the earth without changing it. This is because when we look at a billiard ball, light waves that bounce off it and come to our eye are so infinitesimal that we are confident that they can't affect the ball in any way. In the quantum world, however, the only way to observe one electron is to bounce another electron (or something equivalent) off it. In this process, the observed electron will be changed. In the words of Chairman Mao, "If you want to taste a pear, you must change the pear by eating it."

652 The Heisenberg uncertainty principle is part of quantum mechanics. It was the German physicist Werner Heisenberg who first fully realized the implications of the nature of observation in quantum mechanics. The principle that bears his name states that because a quantum object cannot be observed without changing it, it is impossible—even in principle—to measure certain things simultaneously. For example, you cannot know exactly both its position and velocity at a given time. The more precisely you know the value of the position, the less sure you are about how fast something is moving, and vice versa.

A precise statement of the Heisenberg uncertainty principle is:

$$\Delta X \Delta V \geq h/m$$

Where ΔX is the uncertainty in our knowledge of the position of the particle, ΔV is our uncertainty in the velocity of the particle, **h** is a number known as Planck's constant, and **m** is the mass of the particle.

653 **The uncertainty principle does not say that it is impossible to make precise measurements in the quantum world.** It simply says that if you choose to measure one thing exactly, you must pay for this knowledge by giving up any hope of obtaining knowledge of something else. In other words, if I want to know the position of a particle exactly, I would make a measurement in such a way that ΔX (the uncertainty in position) would be zero. In order for the uncertainty principle to be true in this case, ΔV (the uncertainty in velocity) would have to be infinite—velocity could have any value whatever. You can measure the position exactly, you can measure the velocity exactly, or you can measure both of them to some compromise level of precision. All the uncertainty principle says is that you can't measure both exactly at the same time.

654 **Because of the uncertainty principle, physicists describe quantum mechanical systems in terms of probabilities.** If you can't tell whether a particle is moving ten feet per second or twenty feet per second, for example, you aren't going to be able to predict where it will be ten seconds from now with very high accuracy. Consequently, you are forced to describe the behavior of the particle in terms of a set of probabilities. In this example, you might say that in ten seconds the particle is most likely to be 150 feet away, but there's a chance it will have traveled only 100 feet, and some other chance that it will have gone 200 feet.

In quantum mechanics, then, everything is described in terms of things called wave functions. As the name indicates, the wave function is a wavelike description of the electron or photon or other "particle." The height of the "wave" at a specific point, however, is related to the probability of finding the particle at that point. Thus, if you have a wave with a hump in the middle that tails off at the two ends, you are saying the particle is most likely to be found in the middle and has a very small probability of being at either of the ends.

655 **Einstein was critical of quantum mechanics.** Most people are aware that Albert Einstein spent his later years as an implacable foe of quantum mechanics, even though he was himself one of the great pioneers in this field. He is supposed to have summarized his objections to the probabilistic aspects of quantum mechanics by saying, "God does not play at dice with the universe." The story is that Niels Bohr, a lifelong friend and colleague of Einstein, got so exasperated by repetitions of his quote that he once snapped back. "Albert! Stop telling God what to do!"

656 Quantum particles sometimes act like waves —take the electron as an example. We usually think of it as being akin to a baseball—a localized blob of matter that we normally think of as a particle. There are many experimental situations in which electrons seem to travel around like little bullets. In certain circumstances, however, the electron can exhibit interference—behavior we associate with waves. For example, if electrons are allowed to hit a screen that has two slits in it, there will be alternating bands of high- and low-intensity aggregates of electrons in back of the screen, just as there would be alternating bright and dark bands if light hit the screen. In this kind of experiment, the electron acts as a wave.

657 Quantum waves sometimes act like particles. There is ample evidence that light is a wave. On the other hand, in the photoelectric effect, it seems to act as a particle. The effect (first explained by Albert Einstein in 1905) is this: when light strikes certain metals, it knocks electrons loose. These electrons start coming out of the metal very quickly—too quickly to measure with all but the fastest modern electronic instruments. The only way to explain this quick emission is to say that, at least in this instance, the light is acting like a billiard ball that collides with the electron and pushes it away instantaneously, rather than acting like a wave that gently washes the electron from its atom.

658 The photon associated with ordinary visible light is about three feet long.

A visualization of a photon.

659 Wave-particle duality—is it a problem or not? In the early part of the nineteenth century, physicists assumed that everything had to be *either* a particle *or* a wave. The behavior of things in the quantum world, therefore, presented them with a quandary. It seemed that whether an electron acted like a wave or a particle depended on the kind of experiment that was done. They called this the problem of "wave-particle duality."

Personally, I don't think wave-particle duality is much of a problem. The existence of the "paradox" simply tells us that we have made an incorrect assumption. We are applying the wrong categories to a new situation, because in the quantum world, everything isn't either a particle or a wave. Electrons and photons are what they are—things that sometimes appear to us as particles and sometimes as waves, but which are in reality a third kind of thing with which we have never had direct experience.

With modern fast electronics, it is possible to set up a situation in which a particle is shot at an apparatus, and, *while the particle is in flight,* choose whether you will do a "wavelike" or "particle-like" experiment after it is too late for the particle to change its mind. When these experiments are done, you get the results predicted by quantum mechanics—wavelike behavior in the wavelike experiment and particle-like behavior in the particle-like experiment. The theory is right, but how can you picture an electron if it behaves like that?

660 Wave-particle duality explains the Bohr atom.
The existence of allowed orbits in the Bohr atom was a mystery when the model was first proposed. We now understand that they are the only orbits for which the wave and particle descriptions of the electron are consistent. An "unallowed" orbit might be one in which the electron, when viewed as a particle, would be stable, but on which the electron "wave" wouldn't fit an even number of times. Conversely, it might be one in which the wave fit, but the particle was moving too fast to stay in orbit. Only when the two viewpoints are consistent— when the particle orbit is stable and

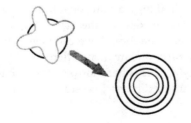

The Bohr atom.

the wave fits—do we get an allowed orbit. Thus, the Bohr orbits are those for which it *makes no difference* whether the electron is a particle or wave.

661 During the later stages of his career Albert Einstein thought of a number of paradoxes by which he hoped to show his colleagues that they had taken the wrong path.
His last sally into this field was done in 1935 when, along with two colleagues, he proposed something that is now called the EPR paradox (the initials come from the names of the authors—Einstein, Podolsky, and Rosen).

Here's what the EPR paradox is: if you have a nucleus that decays into two identical particles, these particles must come off back to back. If the particles are spinning, and if the one moving to the right is spinning clockwise, then the one moving to the left must be spinning counterclockwise. What if you let each of the two daughter particles travel a long distance without being measured and then measure only one of them—the one on the right, for example? If this particle is spinning clockwise, the one on the left must be spinning counterclockwise and, Einstein argued, both particles must have had these spins all along, whether they were measured or not. He argued that this proved that quantum particles "really" had definite properties all along, and that the uncertainty principle was just a result of our inability to measure them. And, of course, if this is true then the real theory that describes the quantum world will have no need of probabilistic explanations.

662 **Bell's theorem makes arguments about the reality of quantum mechanics an experimental question.** In 1962, Scottish physicist John Bell pointed out that by looking at quantities which could be measured (such as the relation between the direction the particle is moving and the direction of its spin) you could test the basic idea behind the EPR paradox in the laboratory. He did this by showing that there were certain relationships between these quantities that would hold if the particle really had a definite spin between the time it was emitted and the time it was measured, but that a different set of relationships would hold if the particle had to be described by a wave function during this transit period (i.e., if, as dictated by quantum mechanics, it has no definite spin when it's not being measured).

During the mid-1970s a number of different laboratories around the world, most notably that of Alain Aspect in Paris, performed the types of experiments that were suggested by Bell. They found that the predictions of quantum mechanics, with its probabilistic electrons and photons, were right—an electron really doesn't have a definite spin until it is measured. Thus, it appears that

nature has simply decided to make the quantum world different from the world we're used to. We may not like it but that is the way things are.

663 **Actually, the most amazing thing to me about the experiments in quantum mechanics is not their outcome, but the fact that they were done at all.** It's almost as if the old sophomore bull-session question—whether a million monkeys with typewriters could produce *Hamlet*—was settled by someone actually getting a million monkeys and setting them to work.

664 **If you want to play the quantum games, you have to play by the quantum rules.** It's one thing to accept the idea that quantum mechanics is probabilistic on an intellectual level. The reason we can do this is because deep in our hearts we believe that the electron *really* is like a baseball, even if physicists want to be fussy about it. Bell's theorem and its experimental outcome force us to accept the fact that the quantum world is fundamentally and irretrievably different from the world with which we are familiar. We have to deal with things at the quantum level through our mathematical equations even though we can never see them or picture them. There's no denying this is hard for people—even hardened physicists—to do.

Special and General Relativity

665 More than a dozen people understand **relativity,** despite the folklore to the contrary. The special theory of relativity is routinely taught to freshmen and sophomore science and engineering students in American universities and colleges, and the essential concepts of relativity are taught to liberal arts majors. Even general relativity, though much more difficult, is studied in a routine way by graduate students in physics and astronomy. Like many other ideas in science, the basic concepts of relativity are simple even if the mathematics gets a little difficult once in a while.

666 Relativity has nothing to do with relativism, the vague statement that "everything is relative." In fact, as next explained, the theory of relativity focuses on those aspects of the physical world that are *not* relative—i.e., on those aspects of the physical world that do not change when an observer changes his or her point of view.

It's a little-known fact that Albert Einstein preferred to call his new theory the "theory of invariants" rather than the "theory of relativity." He felt that that term better reflected his thoughts. Had people listened to him, we might have avoided some of the confusion that has surrouned relativity since its inception.

667 The principle of relativity states that the laws of physics are the same in all frames of reference, and that no matter where you are in the universe, you will discover the same laws of nature operating. This is true even though an event may not look the same to different people. The principle tells us that it is the laws of nature and not the phenomena themselves that are the bedrock of the physicist's view of the world.

For example, if you are standing still and someone going by in a train drops a baseball, the two of you will give different descriptions of what you see. The person on the train will say the baseball drops straight down, while you will say that the baseball moves in the direction of the train while it is falling. The two of you will therefore disagree on your descriptions of the phenomenon, and in this sense, the description of the fall of the baseball is indeed relative.

On the other hand, if you and the other person on the train perform enough experiments to deduce the law that governs falling bodies, each of you will discover exactly the same law. The laws are fixed, the phenomena are relative—this is the general idea that lies behind Einstein's theories.

668 There is a difference between special and **general relativity.** The former is a theory that Einstein first published in 1905. It says that the laws of nature are the same for all observers whose frames of reference are moving with constant velocity with respect to each other. The person on the ground and the person on the train in the previous item are examples of this type of observer, since the train moves by at a constant velocity.

General relativity holds that the laws of nature are the same for all observers even if they are accelerating with respect to each other. Thus, general relativity includes special relativity, but encompasses much more.

669 Relativity was the most famous thing that Albert Einstein did, but it's not what he got his Nobel prize for. In fact, such was the conservatism of the physics establishment in the early part of this century that Einstein's prize was given for his work on the photoelectric effect. Apparently at that time a whiff of heresy still clung to the whole notion that you could learn something about nature by thinking about frames of reference.

670 Relativity does not contradict Newton's mechanics. Both of these theories make predictions about the outcomes of experiments. These predictions differ, but do so significantly only for objects moving very close to the speed of light. For objects moving at normal speeds, the predictions of special relativity and those of Newtonian mechanics are, for all intents and purposes, identical. For this reason, we say that special relativity encompasses Newtonian physics rather than replaces it. It reproduces Newton at low speeds, but describes phenomena more accurately at higher speeds.

671 According to relativity, the speed of light is special. This speed, usually denoted by the letter, "c," has a special role in relativity because it is built into Maxwell's equations. It is the only speed with this distinction and therefore the only speed on which all observers have to agree if the principle of relativity is to be correct.

672 The predictions of relativity do not match up with our everyday experience. For example, if you stand on a railroad car that's moving thirty miles an hour and throw a baseball forward at the speed of twenty miles an hour, you expect that someone on the ground would see the baseball moving at fifty miles an hour—your speed plus that of the railroad car.

Suppose instead that you were on that same railroad car and sent the beam of a flashlight in the forward direction. You would see the speed as being 186,000 miles per second. Someone on the ground, however, would also have to see the speed of the light as 186,000 miles per second—*not* 186,000 miles per second plus thirty miles per hour—if the principle of relativity is to be right. If the person on the ground saw a different speed from you, then Maxwell's equations wouldn't be the same for both observers and the principle of relativity would be wrong.

It is only because relativity is so well verified experimentally that physicists are now willing to accept this sort of strange proposition.

673 According to relativity, moving clocks slow down.

You can imagine making a clock that works as shown in the sketch. A flash bulb goes off, light travels to a mirror, reflects off the mirror and is counted. The entire sequence—flash, bounce, click— would be one "tick" of a clock. If you were watching a clock like this go by on a railroad car, it would look to you as if the light were traveling a sawtooth pattern as shown in the sketch—as the light was moving from the flash to the mirror and back the entire apparatus would be moving to the right. The light in the moving clock would be tracing a diagonal path, while the light in the clock on the ground would simply be going up and down. If the light is moving at the same rate of speed

in both cases, the "tick" on the ground must be shorter than the "tick" for the moving clock. This is the basis of the claim that in relativity moving clocks slow down.

674 The twin paradox isn't really a paradox.

This paradox arises because, according to relativity, if one of two identical twins spent his life in a rocket ship traveling near the speed of light, when he comes back to earth he will be younger than his sibling. Today we know that the twin paradox is a real effect (see the following). In other words, it shouldn't be called the "twin paradox" but the "twin effect."

675 The slowing down of moving clocks can be tested experimentally.

In the 1960s a group of scientists at the University of Michigan put atomic clocks on airplanes that were flying around the world (it was Pan American flight #1, if you must know). After the clocks completed their journey, they were compared to identical clocks that had been left in

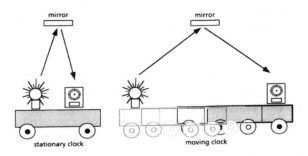

The "light clock."

the laboratory. The result: the moving clocks had, in fact, ticked fewer times than the stationary ones. Of course, these are clocks that can measure time to an accuracy of thirteen decimal places, not your standard wrist watch, but they establish the principle that time is relative.

676 In addition to the slowing down of clocks, special relativity predicts that moving yardsticks will appear to be shortened in the direction of motion, that moving objects will appear to be heavier than they would be if they were sitting still, and that mass and energy are equivalent.

677 The statement "Relativity says that nothing can travel faster than the speed of light" isn't strictly true. What relativity says is this: if you take an object and start accelerating it, it gets more massive. The closer you get to the speed of light, the more massive it gets until, at the speed of light itself, the mass becomes infinite. Since it takes an infinite force to accelerate an infinite mass, the conclusion is that you can never get the object to the speed of light.

However, what the argument actually proves is that nothing now moving slower than the speed of light can be accelerated to that speed. Since photons, by definition, travel at the speed of light, it can't be true that "nothing" can go that fast.

678 Some scientists have suggested that there may be a class of particles that always move faster than light, and cannot be decelerated to lower speeds. They call these particles "tachyons" (from the same Greek root that gives us "tachometer"). No searches for these particles have been successful, so at present we do not know whether they exist or not. If they did exist, they would provide a way of communicating instantaneously with any point in the universe.

Tachyon scientists, by the way, refer to normal particles as "tardyons."

679 $E = mc^2$ That is, the energy within an object equals its mass times the speed of light squared.

680 The equivalence of energy and mass is well verified experimentally. There is a good chance, for example, that the electricity in the light bulb you are using to read this book was generated by the conversion of mass into energy in a nuclear reactor somewhere. So you might say that the fact that your light goes on when you flip the switch is in itself an experimental test of the concept of relativity.

Similarly, in major accelerators around the world, particles are routinely brought to speeds near that of light and then allowed to smash into each other. In these collisions some of the energy of the moving particles is converted into mass and copious numbers of new particles are created where none existed before.

681 General relativity is our current best attempt at a theory of gravitation. The mathematics of general relativity is much more complex than that of special relativity. Nevertheless, the results of general relativity follow from the postulate of general relativity with the same remorseless logic that the results quoted follow the postulate of special relativity. General relativity is our best theory of gravitation to date.

Chemistry

682 Chemistry is the branch of science concerned with the formation and properties of molecules, which are combinations of two or more atoms. Chemistry is the central science. It touches on almost everything else—we study chemical interactions in stars, in minerals, in cells, and in the molecules of living systems, to mention just a few examples.

683 When atoms come together, some reactions release energy, while others require energy to make them go. For example, when oxygen combines with carbon in a wood fire, heat is released. On the other hand, you can't bake a cake unless you put energy in the form of heat into the dough. If the reaction gives off energy, we say it is exothermic, while if it requires energy, we say it is endothermic.

684 Almost every material you encounter in everyday life is made from a combination of atoms, rather than being made of single atoms. Even the air you breathe is made primarily from oxygen and nitrogen molecules (each made from two atoms bound together). The clothes you wear, the food you eat, and even your body are all made from molecules.

685 All molecules are made from around 100 chemical elements. When chemistry first began in its modern form in the eighteenth century, chemists quickly learned that they could break ordinary materials down. A complex structure like wood could be burned to yield carbon (in the form of charcoal) and various gases. When chemists carried out these sorts of operations, they found that there was a fundamental unit beyond which the matter could not be broken down by techniques available to them at that time. These "irreducible" constituents of matter they called elements. Today the roster of chemical elements exceeds 100 items.

686 Each chemical element corresponds to a different kind of atom. Different chemical elements have different numbers of protons in their nucleus and different numbers of electrons in orbit. Thus, an atom corresponds to the element carbon if it has six protons in its nucleus and six electrons in orbit, while it corresponds to the element oxygen if it has eight of each.

687 The propensity of an atom to interact with its fellows depends on the way its electrons are arranged. When two atoms come near each other, each first sees the other's outermost electrons. Whether the two atoms will lock together or go their separate ways depends on the way those electrons are arranged. Thus, it is the outer electrons that determine whether the atom will take part in chemical reactions.

The outermost electrons of an atom are called the valence electrons. Another way of stating this idea, then, is to say that the chemical properties of an atom depend on its valence electrons.

688 Elements heavier than uranium are made in laboratories by scientists who are specifically trying to create them. They tend to be radioactive, and to have short half lives. Most of the last row of the periodic table (see the following) was created artificially. The basic technique for making new elements is to put one nucleus into an accel-

erator, speed it up, and let it collide with another. In the resulting reaction, the two nuclei occasionally fuse together to form a heavier one, which, one hopes, can be identified before it decays.

689 Chemical elements have been given names for a variety of reasons. For example, hydrogen bears its name because it is the generator of water. It is fortunate from the point of view of aesthetics that the German equivalent of this word (*Wasserstoff*) was snubbed for its Greek cognate.

Other elements have been named for the color of the light they emit—cesium means "sky-blue," for example. More recently, chemical elements have been named to honor famous scientists (Einsteinium and Mendelevium) and places (Berkelium, for Berkeley, California, where this particular element was first seen).

690 In 1982, a group at Darmstadt, Germany, succeeded in creating one atom of element 106. This remains the heaviest element ever seen, and that one fleeting atom remains as the world's complete inventory.

The Periodic Table of the Elements

691 The Russian scientist Dimitry Mendeleyev (1834–1907) discovered the periodic table of the elements. He found that if you list the elements in order of increasing weight and then arrange them as shown—two elements in the first row, eight in the second and third, eighteen in the next, and so on, a rather strange regularity becomes apparent. If you read down the columns of this display, the chemical elements all have similar chemical properties. For example, hydrogen, lithium, sodium, potassium, and so on (all the elements in the first column) are extremely reactive chemically—they *like* to combine with other atoms. On the other hand, the elements helium, argon, neon, etc. (the elements in the last row) are all very stable and reluctant to enter into chemical combinations.

When Mendeleyev put the table together he found that there were two gaps—holes where elements ought to be but weren't. These holes corresponded to the elements we now call Scandium and Germanium. When they were discovered, the predictive power of the periodic table had been established.

692 **The structure of the periodic table is a reflection of the underlying laws of quantum mechanics.** In particular, it appears to be a manifestation of something known as the Pauli exclusion principle. This principle says that no two electrons can occupy the same state. If you imagine building the periodic table by adding electrons one by one to the orbits around the nucleus, then there are a limited number of places to put electrons in each orbit.

There are, for example, only two available places in the innermost orbit. Thus, if you want to add a third electron, it cannot go in the lower orbit, but must go in the next higher one. The next orbit can accommodate eight electrons, as can the next, and so on. In fact, there is an exact correspondence between the number of spaces for electrons in an orbit and the number of elements in the corresponding row of the table. You can think of building the table by filling electron orbits (or "shells," to use the chemist's lingo), then starting a new row after each shell is filled.

Atoms in the same column in the periodic table have the same number of electrons in the outermost shell, and therefore have similar chemical properties.

Chemical Bonds

693 **Atoms like to have closed electron shells** —that is, they like to be in a situation where every orbit is filled up. For example, an atom like sodium, which has one electron outside a filled shell, "wants" to give that electron away. In the same way, chlorine, which has seven electrons (one less than the number needed to fill the orbit) in its outer shell, "wants" to acquire an electron to fill the shell.

694 **In ionic bonding, one atom gives up an electron and another accepts it.** For example, in the formation of table salt (sodium chloride), sodium gives up the electron and chlorine takes it. Because of this permanent transfer of an electric charge, the two atoms involved become ions—that is, each has an electric charge of its own. Thus, there is an electrostatic attraction between these two atoms. It is this attraction that binds them together and which, ultimately, pulls the material together. This arrangement is called an ionic bond.

In general, ionic bonds appear in inorganic materials and hold things like rocks and crystals together.

695 **In the covalent bond, one electron is exchanged rapidly back and forth between two atoms.** In effect, the atoms share the electron, and this sharing process serves to hold the atoms together.

The most common atom that forms covalent bonds is carbon, which has four electrons outside of a filled shell. Almost all of the bonds that hold organic materials together are of this type. The tissues in your body are held together largely by covalent bonds.

696 **A vestigial type of ionic bond, called the hydrogen bond, is important in many materials.** It works like this: when hydrogen and oxygen come together, they normally form a covalent bond. The oxygen, however, attracts the electrons so strongly that it tends to pull them toward itself, leaving the positive part of the hydrogen atom somewhat exposed. This exposed positive charge can then exert an electrical force and attract other atoms.

The effect of the hydrogen bond is seen most readily in water, where the molecule has positively and negatively charged ends. This arrangement allows water molecules to exert an electrical force on other molecules, even though the water has no net electrical charge. This is why it is so easy for water to attract other kinds of molecules, and why water is the universal solvent.

697 **In a metal, all the atoms hook themselves together by a single cooperative effort,** with each atom contributing one or more electrons to the general structure. These electrons float loosely throughout the material, so a metal is an array of heavy positive ions in a sea of loose electrons. In essence, you can think of the metallic bond as being the logical extension of the covalent bond—it is a bond in which all the atoms of the material share all the electrons, rather than having the sharing go pair by pair.

698 **van der Waal's bonds are the weakest link in the molecular chain.** Named for the Dutch physicist Johannes D. van der Waals (1857–1918), the van der Waals force is generated this way: when two atoms approach each other, the electron cloud in one atom tends to repel the electron cloud in the other. In effect, this repulsion pushes the electron cloud away from the nucleus in each atom. The result of this mutual distortion is a weak electrical force between the two atoms.

699 Whether a material is hard or soft, flexible or rigid, depends on what kind of bonds go into keeping it together. For example, wood is a fairly rigid material because it is made of long strings of cellulose held together by the hydrogen bond. You can stand on a rock and not fall through because the rock is held together by ionic bonds, one of the strongest interatomic forces. You can crumble clay in your hand because the force between adjacent sheets of molecules in clay is of the van der Waals' type. No matter what the property of the material is, you can understand it in terms of the way that the atoms are held together.

700 Whether a material appears as a solid, liquid, or gas depends on the way its molecules are held together. The types of chemical bonds we've just discussed appear primarily in solids, where they tend to lock atoms into a rigid structure. If you push on one atom, the force is transmitted to all the others, and the entire solid moves.

In a liquid, on the other hand, molecules are packed close to each other but are not locked together—they're like marbles in a bag. Pushing on one molecule does not force the others to move.

In a gas, the molecules are widely separated and move around like billiard balls on a table—occasionally colliding but generally interacting only minimally with each other.

701 Plasma can be thought of as the fourth state of matter. When the temperature of a gas is raised high enough, the collisions between atoms become violent enough to knock electrons loose from their nuclei. These collisions result in a gas made up of loose negatively charged electrons and heavier positively charged nuclei. This state of matter is a plasma. Plasmas are found (among other places) in fluorescent light bulbs and in stars.

Organic Chemistry

702 Organic chemistry began as the study of molecules found in living systems. At one time, chemists believed that molecules found in living systems were different from those in nonliving systems. When they finally succeeded in synthesizing complex molecules in the nineteenth century, however, this point of view changed, and it was recognized that the same laws applied alike to living and nonliving matter. Today, the term "organic chemistry" is generally used to refer to the study of compounds that contain carbon and hydrogen, whether they come from living systems or not. Thus, someone trying to make synthetic gasoline would be considered to be an organic chemist, even if he or she does not necessarily start with materials taken from living systems.

703 Organic chemistry depends on the unique features of carbon, which has four electrons in its outermost orbit. These electrons can link carbon, by way of the covalent bond, to other atoms. What makes carbon important is the ability of carbon atoms to link to each other and form long chains. These carbon chains form the basis for the molecules that make up all living systems on earth.

Two carbon atoms can link together either by exchanging one electron, in which case we say there is a single bond, or by exchanging two electrons, in which case we say there is a double bond. Obviously, the number of bonds any other atom can form depends on the number of electrons in the outer orbit.

704 The most important aspect of organic chemistry is the three-dimensional structure of the molecules for this is what determines whether molecules "fit together" in the ordinary geometrical sense so that the bonds between relevant atoms are allowed to form. Two molecules that would be capable of locking together may not do so if they are not oriented correctly. You can think of the atoms as being analogous to complex structures that have little spots of Velcro on them. If the alignment of the molecules isn't just right—if the Velcro pieces don't come together—then the molecules will not stick together.

705 There is a code to help us in drawing organic molecules, because when the number of atoms in a molecule gets to be twenty, thirty, or more, we could get completely lost in the details. The standard notation used by chemists for sketching molecules is as follows:

1. A bond is represented by a straight line between two atoms.

2. If the atom in question is carbon, it is not shown specifically in the picture.

3. Hydrogen atoms are not shown at all.

In this kind of notation, what would be a complex drawing becomes very simple. For example, on the left, I show a drawing of the molecule glucose, one of the basic sugars, with all the atoms present.

Two ways of drawing the glucose molecule.

On the right, we show the same molecule as it would be drawn using the rules just given. In the rest of this book, we will use this particular way of drawing molecules.

Note to reader: More references to chemistry are to be found in the sections on "Molecular Hall of Fame," "Rocks and Minerals," "Molecules of Life," and "The Genetic Code."

A Glossary of Chemical Terms

One of the great problems that people have in studying chemistry is the large number of special terms that chemists use. In this section, I give short definitions of some that have not been discussed elsewhere in the book.

bon-oxygen arrangement shown. Various esters differ in the types of atoms in their end group. Many fragrances are carried by esters—when you sniff a rose, for example, it is esters that trigger your sense of smell.

An ester.

Terms Referring to Molecules

706 **Acid:** Any molecule that donates a proton to other molecules in a chemical reaction. An acid is the opposite of a base. Very strong acids, like the liquid in your car battery, are highly corrosive.

707 **Base:** Any molecule that accepts a proton in a chemical reaction. It is the opposite of an acid. Strong bases, like lye, are also highly corrosive.

708 **Esters and Polyesters:** An ester is a molecule that has the characteristic structure sketched. It consists of one group of atoms linked to another by the car-

A polyester is a molecule made up of a string of individual esters linked through the same carbon-oxygen structure. Polyesters tend to be long thin filaments and hence are used widely in making artificial materials like Dacron. There is a reasonable chance that you are wearing something made of polyester as you read this.

709 **Polymer:** Any molecule formed by connecting smaller molecules together, either in a string or in some more complex shapes. Proteins, starches, cellulose, and PVC are all examples of polymers.

710 **Polypeptide:** A polymer that is made by stringing amino acids together. Polypeptides are a special sort of polymer. Proteins are examples of molecules that are both polypeptides and polymers.

711 **Saturated and unsaturated bonds:** We say that a particular molecule is unsaturated if there are double bonds between carbon atoms in it. In this situation, two carbon atoms will be sharing two electrons, and it is relatively easy for an outside atom to come in and "steal" one of them. A molecule is polyunsaturated if it has more than one double bond in its carbon chains. Conversely, a molecule is said to be saturated if all of the bonds between carbons are single. In this case, it takes a fair amount of energy to break into a bond and initiate a chemical reaction.

You are most likely to run into the terms "saturated," "unsaturated," and "polyunsaturated" in looking at food labels. Nutritionists generally hold that unsaturated or polyunsaturated fats are better for you than saturated fats. The term "hydrogenated" is often used on labels to describe a molecule which was originally unsaturated but to which hydrogen has been added to take up the double bonds. This is done to improve shelf life of the material, but can have unfortunate dietary consequences.

Terms Referring to Reactions

712 **Catalyst:** Any molecule that facilitates a chemical reaction between other molecules or atoms but which is not itself involved directly in the reaction or changed by it. One catalyst with which you are probably familiar is the platinum in the catalytic converter in your car, which facilitates the removal of pollutants from unburned gasoline.

An enzyme is a catalyst for reactions involving complex organic molecules.

713 **Distillation:** A method for separating a mixture of two liquids that have different boiling points. For example, if you heat a mixture of alcohol and water to a temperature just below 100 degrees Celsius, the alcohol will boil but the water will not. The vapor that comes off the liquid will have proportionally much more alcohol in it than does the original liquid. In this case, distillation is used to concentrate alcohol. "Distilled spirits" like whiskey are made by distillation.

Crude oil is also processed by distillation, and different kinds of petroleum products (gasoline, benzene, etc.) are taken from the crude oil by a series of distillation processes.

714 **Oxidation:** Any chemical process that removes electrons from a molecule. The most common chemical reactions in which this occurs involve combination with oxygen (hence the name). The burning of wood is an example of oxidation.

Today chemists use the term "oxidation" in a more general sense, even to the point of using it to describe reactions in which no oxygen is present.

715 **Reduction:** The opposite of oxidation, reduction describes reactions in which electrons are added to molecules. Historically, the term referred to reactions in which oxygen was removed and hydrogen added in its place. As with oxidation, the term "reduction" is now used more generally, and may refer to reactions in which no hydrogen or oxygen is present at all.

Terms Referring to Mixtures

716 **Alloy:** A metal made from the mixture of two or more other metals (or of a metal and nonmetals). Brass, for example, is an alloy of zinc and copper.

717 **Colloid:** Small bits of material (larger than molecules) suspended in a liquid. The material doesn't dissolve, but the particles are so small that the entire system acts like a fluid.

718 **Emulsion:** A mixture of two or more fluids in which one of the fluids exists in the form of tiny droplets or particles within the other. Milk is an example of an emulsion, as are some salad oils.

Molecular Hall of Fame

Useful Molecules

719 **Isooctane:** This particular molecule is typical of the sorts of things you put in your car's gas tank. It contains eight carbon atoms (hence the "oct"). What we call "gasoline" is actually a mixture of many different kinds of molecules such as octane. Some of these molecules (such as ordinary octane) are in straight carbon chains. Isooc-

tane is branched, as shown, and is considered the most desirable fuel because it burns smoothly.

In this drawing, as in those that follow, we are using the chemist's notational scheme discussed in CHEMISTRY.

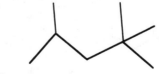

Isooctane.

720 A fluid composed completely of isooctane is given an octane rating of 100, while one made of a different hydrocarbon called heptane (seven carbons in a straight chain) is given an octane rating of 0 because it is considered to be a very undesirable fuel (it causes knocking). A 90-octane fuel is equivalent to a mixture of 90 percent isooctane and 10 percent heptane.

721 Polyvinylchloride (PVC): Like many large molecules, polyvinylchloride is modular—it is a long repeating string of smaller subunits. Polyvinylchloride is widely used for all sorts of purposes, including water pipes and other plumbing fixtures.

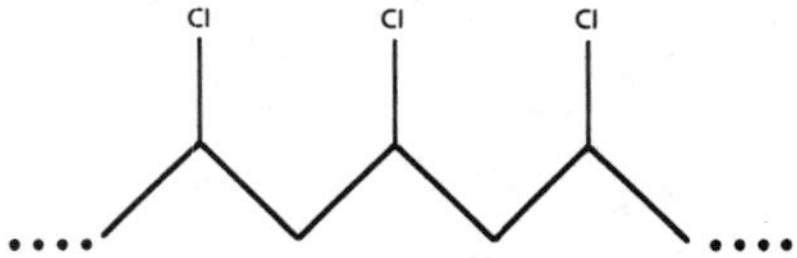

Polyvinylchloloride.

722 Morphine: Morphine fits into a receptor on a nerve cell and hence is a natural pain killer. The addictive drug heroin is a derivative of morphine.

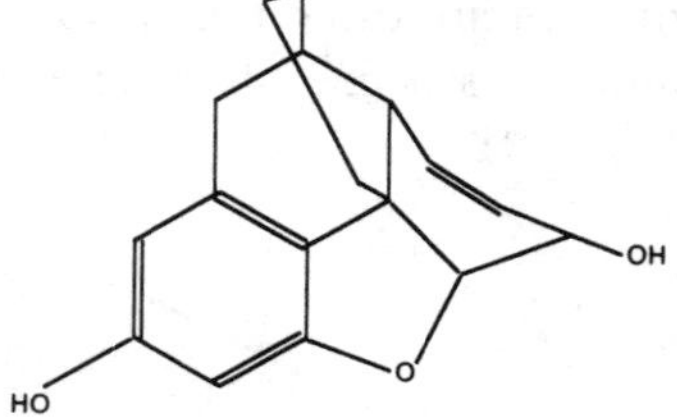

Morphine.

723 Salicylic acid: The principle component of aspirin, it apparently relieves pain by blocking the formation of a neurotransmitter.

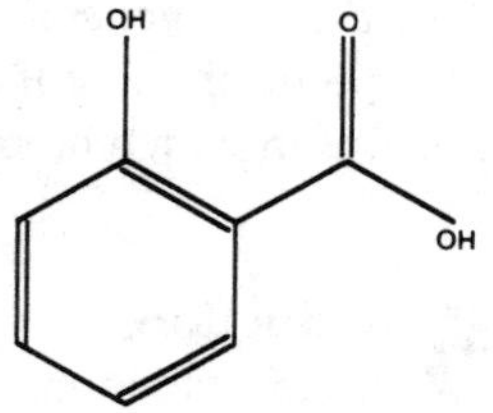

Salicylic acid.

Bad Guys

---

724 Trinitroltoluene (TNT): The principle explosive in conventional bombs. It is an explosive because the oxygen atoms on the three groups shown are almost unstable. Given the slightest provocation, they will leave the atom, carrying with them

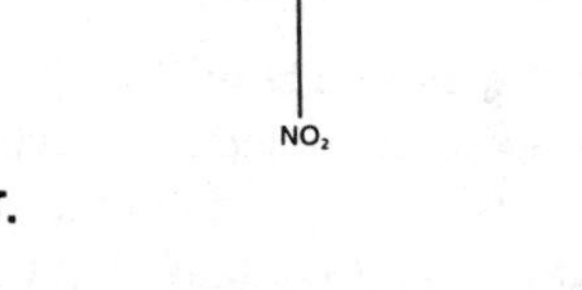

TNT.

some of the hydrogen and carbon atoms on the inside. In effect, the molecule burns itself, producing carbon dioxide and water. The explosive properties result from the fact that the molecule converts a very small amount of solid into a large amount of gas under high pressure, a process that leads to a rapid expansion.

725 Cholesterol: Although cholesterol has gotten bad press these days, it is produced in the body and is used in the production of cell membranes. Unfortunately, it also contributes to the buildup of sludge in the arteries and to atherosclerosis.

Cholesterol.

Molecules You Eat

726 Ethanol: The active ingredient in alcoholic beverages. Its primary effect comes from the fact that it is the right shape to be recognized by receptors on nerve cells. When alcohol binds to these receptors, it changes the shape of channels that allow chemicals into the cell, and therefore affects the working of the nerve.

A number of tranquilizers act in exactly the same way. This is why it is often very dangerous to take tranquilizers and alcohol at the same time.

Ethanol.

727 Saccharine: One of a large number of artificial sweeteners, saccharine tastes sweet because it binds to protein receptors in the taste buds of the tongue in roughly the same way that sugars do. However, it is not metabolized by the human body and therefore adds no calories. Saccharine was discovered in 1879 and has been in production since the beginning of this century.

Saccharine.

728 Caffeine: This substance is the stimulant in coffee and tea. Its effect on the human body is somewhat circuitous. What it does is pass itself off as ATP, the universal energy currency in cells, thereby blocking the production of enzymes that inhibit ATP production. The effect of the caffeine, then, is to increase production of ATP. A close relative of caffeine, called theobromine, is the stimulant in chocolate.

Caffeine.

Important Molecules in Biological Processes

729 **Glucose:** The natural sugar that carries the energy in most biological systems, it appears in two forms—as a straight line, as shown on the left, and as a ring, as shown on the right. Note the similarity of the structure of glucose to that of isooctane, the principle component of gasoline (see the foregoing). Both of these are fuels, and both of them are important to us because of the energy they carry.

730 **Chlorophyll:** The central molecule in photosynthesis. It is chlorophyll, of course, that makes leaves green.

731 **Adenosine Triphosphate:** ATP is the universal energy coin in workings of the cell. It can store energy because it requires energy to add the last phosphorus-oxygen group onto the tail of the molecule. Thus, if you have energy available, you can push one of these groups on and form the ATP molecule. The molecule then travels to other parts of the cell where the group can be taken off, releasing the stored energy in the process.

732 When an animal dies, ATP is no longer synthesized in the cells. The muscles, no longer capable of using energy, stiffen. This explains why if someone dies after a fight or after being extremely frightened (processes which deplete the supply of ATP), rigor mortis sets in more quickly. This bit of trivia is well known to aficionados of murder mysteries.

A Small But Important Difference

733 **Testosterone and estradiol:** The two molecules shown on the facing page look very similar. They differ only in the structure of the left-hand ring and the hydrogen (or OH) that is attached to it. Yet that small difference in the molecules makes all the difference in the world. The molecule shown on the left is testosterone. This is the principle male sex hormone. The secretion of testosterone starts at puberty and controls all the secondary male sex characteristics in humans.

The molecule shown on the right is estradiol, the principle female sex hormone in humans. Secretion of estradiol begins at puberty and

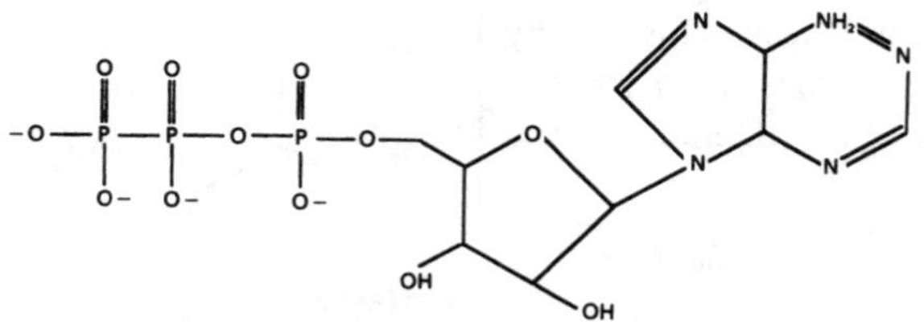

ATP.

Estradiol (left) and testosterone (right).

phases out at menopause. It controls the secondary female sex characteristics.

This example shows that you don't have to make much of a change in a molecule to make a very large change in the organisms in which that molecule is found.

6

EARTH SCIENCE

Earth.

The Formation of the Earth

734 **The earth was formed at the same time as the sun.** When the sun condensed out of a cloud of interstellar gas, a small amount of material was left rotating in a disc outside its main body. Our current theory is that in this disc the gravitational force operated to bring materials together into what astronomers called planetesimals—chunks of rock and frozen liquids ranging in size from a a few feet to a few miles across. These planetesimals began to come together to form the planets, including the earth.

735 **While the earth was forming, it was heating up and becoming differentiated.** Every time a planetesimal joined the newly forming earth, its kinetic energy was converted into heat, and the net effect of these impacts was to melt the newly forming earth. During this heating phase, heavy materials (like iron) sank toward the center of the earth while lighter materials (like silicon minerals) floated up toward the top. Like the ingredients of a salad dressing that has been allowed to sit too long, different materials in the earth separated from each other. Geologists say that the earth became "differentiated" during this early phase of its existence.

736 **Radioactivity produced heat in the** new earth. The gas from which the solar system formed contained a certain complement of radioactive nuclei. As these nuclei were incorporated into the earth, they continued to undergo radioactive decay, generating heat as they did so. The difference between radioactive heating and heating by impact is that the bombardment largely stopped once most of the loose material in the neighborhood of the earth had been incorporated. Radioactive heating, on the other hand, goes on today and will continue to do so until all the unstable nuclei have decayed.

737 **The earth has a layered structure,** as shown. At the center are the the heaviest materials, mostly nickel and iron in a structure called the core. There is a solid inner core a little less than 800 miles in radius surrounded by a liquid outer core extending outward another 1300

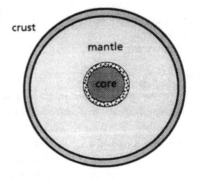

The interior structure of the earth.

miles. Above the core, 1800 miles thick and extending almost to the surface, is the mantle—a region of the earth made largely of solid rock. Finally, the outer part of the earth—roughly the last thirty miles—is made of lighter rocks and is called the crust. Both the continents and the ocean floors are part of the crust. The earth's materials have been segregated according to their density, with the heaviest materials located in the center and the lightest at the top.

738 **The core has both a solid and a liquid part because both temperature and pressure increase as we go deeper into the earth,** so that at the lower pressure in the region of the outer core the iron-nickel material can still be a liquid, but farther in it is compressed into a solid.

739 **If the earth had been perfectly differentiated, there would be no iron or heavy metals in the crust at all.** In fact, the process of differentiation was not total, so there are traces of heavy metals left at the earth's surface. It is these traces that we mine for the metals we use.

740 **The farther down you go in the earth, the warmer it gets,** and, although the details vary from one place to another, the general rule is that once you get more than a few hundred feet beneath the surface, the rocks heat up several degrees celsius for each 1000 feet of depth. This is why deep mines (such as gold mines) are such uncomfortable places to work—it's not at all unusual for rock temperatures to exceed 130° F when new shafts are opened.

741 **Heat flows out from the earth** because the interior is hot. This flow amounts to only 2 percent of the energy that comes from the sun, so it has little effect on living systems. It is, however, extremely important in geological processes. The amount of energy that leaks out of a square meter of the earth's surface (on the average) is enough to run two TV sets continuously.

Enduring Puzzle

742 **Where does the heat come from?** There are two sources of heat in the interior of the earth. One is the original heat that was generated when the planet formed, the other is radioactivity. Geophysicists disagree over how much of the heat in the earth comes from each of these sources. Is the heat flow mostly energy left over from the early heating or is it mostly the result of radioactive decays? Until we know more about the details of the materials in the planet's interior, we won't know.

743 **Using devices called diamond anvil presses, scientists can produce temperatures and pressures in excess of those found at the**

center of the earth. The technique involves putting a sample of material between two pieces of diamond, then squeezing down. Since diamond is transparent to light, laser beams can be directed onto the material in the press when it is under high pressure to raise its temperature. In the late 1980s, scientists finally were able to study samples of materials at the temperatures and pressures that are found at the center of the earth.

744 **The deepest test boring made anywhere on the earth is in the Kolyma peninsula of Siberia,** where the Soviet Union has a geological test site. At last account, the drill bit had passed 10 kilometers (6 miles), which is pretty close to the limit of what is technically possible.

745 **Most of our knowledge of the earth's interior comes from monitoring seismic waves.** Whenever there's an earthquake, waves move out through the rocks in which the earthquake occurs. Waves of this type are called seismic waves. Some of these waves travel through the interior of the earth and can be detected by suitably equipped laboratories all around the surface. By noting where they reach the surface and how long it takes them to get there, scientists can get a pretty good idea of the properties of the material through which they have traveled. Most of the general structure of the earth was deduced from this sort of data.

Evolution of the Atmosphere and Oceans

746 **The earth was born without atmosphere or oceans.** If there was water vapor and atmospheric gas on the original earth, intense solar winds from the early sun would have blown them away. Therefore, for all intents and purposes, the earth started as a solid ball of molten rock without atmosphere.

747 **The earth acquired its atmosphere through "outgassing."** Each time a volcano erupted or a new hot spring formed, gases from the crust and upper mantle were brought to the surface. This collection of gases became the earth's first atmosphere. When temperatures dropped below the boiling point of water, water condensed from this atmosphere to form the earth's first oceans.

748 **The composition of the earth's early atmosphere was very different from what it is now.** The conventional theory is that the early atmosphere was made of methane, ammonia, carbon dioxide, and water—no oxygen or nitrogen. From the molecules in this atmosphere, the first living systems were made.

749 **The first living systems on the earth were probably blue-green**

algae. These algae would have used the carbon dioxide in the atmosphere together with sunlight to run photosynthesis, giving off oxygen as a waste product. At the same time, sunlight was breaking up water molecules in the upper atmosphere, freeing oxygen. The net effect of the addition of oxygen to the atmosphere was a shift in its composition, a shift called "the great turnover." This occurred about 2 billion years ago. At this time the earth went from its primitive atmosphere to something very like what we have today—an atmosphere rich in oxygen.

750 **The total amount of water at the surface hasn't changed much since the beginning.** Most of the water now in the earth's oceans came out of the atmosphere in that first rainfall. Today the earth loses roughly a swimming pool full of water every year to space, and acquires about the same amount from water seeping up at mid-ocean vents. Thus, the water that you use today is the same water that has been used by every other living thing in the history of our planet.

751 **The moon is made from materials that have roughly the same density as the mantle of the earth.** But the moon has almost no iron such as that in the earth's core. This fundamental difference poses an enormous problem for scientists trying to understand how the moon evolved.

752 In the days when the fission theory was riding high, the Pacific basin was thought to be the hole from which the moon had been torn—its "birth scar." Today, of course, due to the discovery of tectonic plates, we know that the Pacific basin is only a temporary feature of the earth's surface.

Enduring Mystery

753 **Where did the moon come from?** There are several different theories about the creation of the moon. The first is the fission theory, which holds that the moon was torn out of the earth at some time in the geological past. This would explain why the moon has a different composition from the earth—it was created from materials in the earth's mantle after differentiation had taken place.

The second class of ideas says that the moon was created elsewhere in the solar system and captured by the earth. These "capture" theories explain how the composition of these two objects can be different, but run into problems when they try to talk about the details of the capture. It turns out to be very difficult for one object to capture another.

A more recent addition to the repertoire of explanations of the existence of the moon is called the big splash theory. In this scenario, a giant meteorite hit the earth soon after it was formed, but after most of the heavy material had sunk to the center. It splattered a lot of material in the upper layers of the earth out into orbit, and this material formed into the body we call the moon.

If I had to make a bet right now, I suppose I would bet on the big splash theory. I wouldn't, however, bet a whole lot.

Dating the Earth

754 **Dating any material is a difficult job—** rocks don't have signs on them saying "Formed 10 million years ago." There are two basic techniques for establishing dates for geological purposes. They are: (1) position (which leads to a relative time scale); and (2) radiometric dating (which leads to an absolute time scale).

755 **In the absence of evidence for disruption, the deeper a rock is buried in the earth, the older it is.** The layers at the bottom of the Grand Canyon, for example, were formed before the layers at the top. This simple principle leads to what geologists call a relative time scale. You can say which layers came first and which later, but you can't say how much later one layer was than another, nor can you say how long it took to lay any individual layer down.

756 **Fossils are used as markers in establishing the relative time scale.** It is often important to be able to cor-relate the rocks in one column with those in another—to answer questions like "Did this layer in the Grand Canyon form before or after this other layer in California?" The correlation between different columns of rock is done by means of marker fossils. The idea is simple: if there is an animal that enjoyed a widespread geographical distribution but lasted only a short time, and if this so-called "marker fossil" is found in two layers of rock far apart from each other, then you can surmise that the two layers were formed at the s ,ne time.

757 **Homo sapiens will make an absolutely splendid marker fossil.** There were no representatives of our species on earth more than a few hundred thousand years ago, and in that short time we have spread ourselves around the globe. If our worst nightmares are realized and we succeed in wiping ourselves out, we'll at least have the satisfaction of knowing that future paleontologists coming across our fossil remains will know precisely when they were formed.

758 **Radiometric dating is based on nuclear half lives.** If you know how many atoms of a given radioactive isotope were present when an object formed, and if you know both the half life and how many of those atoms are still around, you can estimate the age of the object. For example, if there were 1000 atoms present at formation, and if there are 500 left, then the object was formed one half life ago.

This technique is known as radiometric dating, and different variations on the theme supply scientists with all of our firm dates from the past.

759 **Carbon-14 dating is the best known of the radiometric techniques.** The idea is that carbon is being constantly taken from the atmosphere and incorporated into living tissue. This carbon contains a known percentage of the radioactive isotope carbon-14. When the living thing dies, the carbon-14 starts to decay. It has a half life of 5730 years, so it is an ideal isotope to measure the time of the death of things that lived during the last several tens of thousands of years. It is particularly important for archeologists, since the time of death of a tree, for example, is a good estimate for the time at which a tool was made from its wood.

760 **Similar techniques are used to date the time of formation of rocks.** One common technique uses the atom potassium-40 (half life 1.3 billion years), which decays into argon-40. Since argon-40 is not incorporated into minerals, all argon-40 atoms in a rock must have come from the decay of the potassium that was there originally. Measuring the argon, then, gives the date at which the rock was formed. This technique was used to date rocks brought back from the moon by the Apollo astronauts.

761 **The age of the earth is about 4.6 billion years.** The oldest rocks are about 3.9 billion years old, and the earth has to be older than that. Moon rocks and meteorites (both formed at the same time as the earth) are both 4.6 billion years old. This age is usually taken to be the age of the earth.

762 **The oldest rock on the earth is almost 4 billion years old**—3.96, to be exact. This is a grain of zircon found in much younger rocks in Canada. The oldest rock formation, in western Greenland, is dated at about 3.8 billion years.

Plate Tectonics

763 **The surface of the earth is constantly changing.** Mountain chains are thrown up and worn down by erosion, oceans appear and disappear, all on time scales of a few hundred million years. Nothing lasts forever. Alone among the planets in the solar system, the earth is still in the process of forming itself.

764 The earth's surface is made from plates in constant motion, a theory that geologists call "plate tectonics." "Tectonics" comes from a Greek word which means "to build"—it's the same root you find in "architect." Plates are structures that cover the earth's surface (see the following), are roughly thirty miles thick, and vary in size from a few hundred to a few thousand miles across.

A force inside the earth, perhaps arising from convection (see the following), drives the motion of the plates that provides the basis for constant change on the surface.

765 Plate tectonics is not the same as continental drift, a theory first put forward in 1915 by the German meteorologist Alfred Wegener. He marshaled evidence supporting a theory that the continents were in motion, but this theory was not the same as plate tectonics. In Wegener's theory (to take one example) the continents not only move, but increase in altitude. Real continents do not get higher as they move.

766 Plates are made from heavy rocks like basalt. The lighter continental rocks (such as granite) float along on top of the basalt like passengers on a raft. Some plates have continental "passengers," others do not.

The North American plate, for example, extends from the middle of the Atlantic Ocean to the West Coast of the United States, ending at California's San Andreas Fault. The Pacific plate, on the other hand, has no continental burden at all—it is a plate that is purely ocean bottom.

767 The driving force that keeps plates in motion is probably convection in the earth's mantle. There is too much heat in the interior to be passed to the outside by the mech-

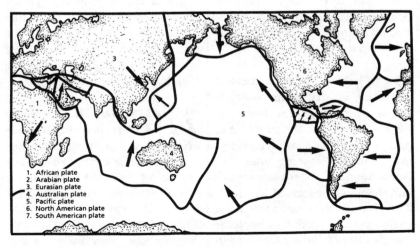

1. African plate
2. Arabian plate
3. Eurasian plate
4. Australian plate
5. Pacific plate
6. North American plate
7. South American plate

Plates at the earth's surface.

anism of conduction, so the rocks in the mantle must undergo convection. Over a period of several hundred million years, the rocks near the center of the earth rise to the top, give up their heat, and sink to the bottom again.

The plates ride along on top of these convection cells, and the continents—dry land—ride along on top of the plates. Looking at the earth's surface, then, is like watching a layer of oil on top of boiling water. Everything is in constant motion in response to things happening deep in the interior.

768 Plate boundaries are where the action is.

Since plates are thick layers of rigid rock, relatively little happens except at places where plates come together—the regions we call plate boundaries. Earthquakes, volcanoes, and other manifestations of geological events tend to be clustered in these boundary regions. There are three major types of boundaries between plates—neutral, diverging, and converging—as shown in the sketch.

769 New crust is created at diverging boundaries,

where hot rock from the mantle wells up to the surface and pushes two plates apart. If a diverging boundary appears under an ocean floor, an underwater mountain chain will be formed. The mid-Atlantic ridge, the longest mountain chain on earth, is one example. It stretches from Iceland all the way to Antarctica. The North American and

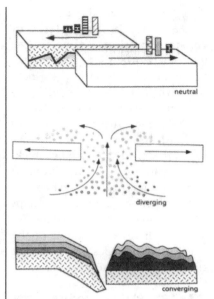

Plate boundaries: neutral (top), diverging (middle), and converging (bottom).

Eurasian plates are being pushed apart along the ridge, and the Atlantic Ocean is getting a few inches wider each year.

If a diverging boundary appears underneath continental rocks, the continent is literally torn apart as the plates separate. This is happening now in the Great Rift Valley of the Middle East and eastern Africa.

770 Surface is destroyed at converging boundaries,

where one plate is pushed underneath the other. We call this process subduction; we say that the plate which is pushed down is being subducted, and call the region where the process is going on a subduction zone. Eventually the material in the subducted plate is melted and its store of atoms returned to the general supply in the earth's interior.

There are several types of subduction zones. If the two plates in question have no continental passengers, the result of a subduction process is a deep oceanic trench. The Marianas trench off the Philippines was formed in this way. If one of the plates has a continental passenger, the continental material will be "crumpled up" as it floats over the subduction zone and a long chain of mountains is formed, perhaps with an adjacent ocean trench. The Andes in South America are an example of this process at work right now. If both plates are carrying continental passengers, the two continents are "welded together" to form a mountain chain. The Ural Mountains mark the spot where Asia and Europe came together and the Himalayas the place where the Indian subcontinent joined Asia.

771 Neutral boundaries are marked by earthquakes as one plate slides along the other. This is happening along the San Andreas Fault in California today, and explains the numerous earthquakes in the San Francisco–Los Angeles region.

772 Plate tectonics presents a unified picture of the operation of our planet, for it shows that all long-term geological processes are related to the motion of plates, and that this motion, in turn, depends on the motion of rocks in the mantle in response to the earth's interior heat. The planet is like a marvelous machine in which all the parts fit together and derive their energy from the same source.

773 Two hundred million years ago, all the land of the earth was collected together in one mass, which we call Pangaea ("all earth"). Then this single mass split into two smaller masses, Gondwanaland and Laurasia. Further splitting brought the continents to the position in which they are now seen. In the future, they will continue to move and the shape of the land will change, as it has in the past.

774 Because of the continental motion, neither ice caps nor rain forest (two things that mark today's earth) have always been present on our planet. There are major ice caps only when a continent is over the North or South Pole. There are rain forests only when the continents are lined up generally in a north-south direction. For most of its history, the earth has not had either ice caps or rain forests, and the climate was very different from what it is now.

775 There is direct experimental proof for plate tectonics. In the mid-1980s, astronomers directed radio telescopes in Europe and the United States to look at the same pulsar and then measured the difference in the time of arrival of radio waves. In this way, extremely accurate measurements of the distance between the two telescopes were obtained, and it was seen that over the course of a year these distances changed by a matter of inches. This affords direct proof that Europe is moving farther away from the United States, as the plate tectonic theorists have been saying all along.

776 **The earth is the only planet that has tectonic activity,** primarily because it is the largest of the terrestrial planets and more heat is generated in its interior relative to its surface than is the case for other planets. Mercury, the moon, and Mars are all small enough so that all the heat that they generate can be taken to the surface by conduction. Venus, almost the same size as the earth, may have had tectonic activity at one time, and may even have active volcanoes on its surface now, but at the moment there appears to be little evidence for plate motion.

777 **In geophysics these days the study of hot spots is a hot topic** (excuse the pun). These are regions where bubbles or plumes of hot material are rising through the mantle, independent of any convection cells. You can think of hot spots as being analogous to the random bubbles you see in a pot of water just before it boils. When the material that creates the hot spot reaches the crust, it pushes the crust up. The idea is that the hot spots are fixed, and tectonic motion drags the plates across them. One result: chains of volcanic islands. The Marianas and Hawaiian islands are thought to represent such chains.

Geological Features of the Earth

Mountains

778 **Mountains are not forever**—they last no more than a few hundred million years. The Appalachians on the east coast of the United States are low, rounded mountains that are near the end of their normal life span. The Rockies, on the other hand, were formed about 65 million years ago and still bear the craggy appearance of newly born mountains.

There is no single "mountain building" mechanism on the face of the earth, and each mountain or mountain chain has its own story to tell.

779 **Fold mountains.** Parts of the Appalachians were formed when tectonic activity caused what is now Europe to collide with the North American continent. As a result, the continental rocks folded like a tablecloth, forming a series of long, parallel ridges interspersed with valleys. This is why roads in places like eastern Pennsylvania tend to run southwest to northeast—they follow the neatly symmetric valleys.

780 **Basin and range.** In Nevada and Utah, mountain chains were formed when large chunks of rock sank and left other chunks standing. The result is a repeating pattern of basin and range, with the edges softened by erosion. This so-called "fault block" process will go on when the surface is put under tension by tectonic forces. The Sierra Nevada range in California is another example of this type of mountain.

781 **Dome.** Sometimes rock is simply pushed up from beneath the earth, more or less like a piston being driven upward. This might happen, for example, if there was a hot spot underneath that particular region. The uplifting produces mountain ranges that are roughly circular in shape and are not connected to any nearby ranges. The Black Hills of South Dakota are an example of mountains formed in this way.

782 **The Rocky Mountains are a very complex geological feature and cannot be attributed to any single cause of mountain building.** Some of them clearly were created when tectonic activity caused small bits of continental material to "dock" in the western United States. Others may be the remains of folding or uplifting activity independent of docking. Working out the precise geological provenance of the Rocky Mountains remains a task that occupies many geologists.

Volcanoes

783 **Volcanoes bring up hot magma from the lower crust and upper mantle,** and tend to occur where the crust is being put under stress. In fact, a map of the earth's active volcanoes bears a striking resemblance to a map of plate boundaries. The "rim of fire," a collection of volcanoes that encircle the Pacific Ocean, is the best example of this phenomenon.

One common driving force for volcanoes at plate boundaries is seen in subduction zones, where heat generated by friction as one of the plates slides underneath the other (as well as by radioactivity in the plates themselves) rises to the surface and creates a chain of volcanoes. The string of islands in the western Pacific from the Aleutians through Japan to the Philippines is a good example of the results of this type of vulcanism.

784 **Central eruptions produce the familiar cone-shaped volcano.** In this process, magma comes up from underneath the volcano in a single fissure and is then spewed out. The cone is built up from ash that falls to the earth and from the magma that cools and solidifies.

The crater formed by a volcanic

Mount Fuji in Japan is a well-known volcano, now dormant.

explosion is called a caldera. Crater Lake in Oregon is an example of a caldera, as is the entire area of Yellowstone National Park.

785 Eruptions can occur in fissures. In its most extreme form, this sort of eruption gives rise to large basalt plains such as the one at the junction of states of Washington, Idaho, and Oregon. Such so-called flood basalts are the primary source of igneous rocks on the surface of the earth.

786 The legend of the lost continent of Atlantis is probably based on the fate of the island of Thera, off the coast of Crete in the Mediterranean. In 1628 B.C., a large volcanic eruption destroyed most of this island, leaving it today as a small rim of rock on one side of a large water-filled caldera. The eruption, along with the attendant tidal waves, has been blamed for the destruction of Minoan civilization.

787 The largest eruption in recent history was at the island of Krakatau in what is now Indonesia. In this 1883 event, several cubic miles of material were blown into the air and an entire island reduced to about a quarter of its former size. The shock from the explosion traveled around the world several times and was recorded in laboratories throughout Europe and the United States. The most recent eruption in North America was Mount Saint Helens in 1980. This eruption, although it caused intense local damage, did not have much effect on the rest of the country.

Glaciers

788 Glaciers are large accumulations of moving ice. It is etimated that glaciers cover 10 percent of the earth's land area, and tie up 5 percent of the earth's water. Glaciers are often located in high mountains, but the main ice sheets on the earth are in Greenland and Antarctica. The ice cap over Antarctica is several miles thick at points.

When snow first falls, it tends to pack very loosely, but as the snow settles, the flakes are broken down and the resulting ice is very densely packed. This so-called "firn" is the principle constituent of glaciers. A glacier will melt at its front face at the same time that ice is accumulating from snowfalls higher up in the mountains. Whether a glacier advances or retreats depends on whether its snow budget is such that more snow is added than is removed each year.

789 **Glaciers flow like rivers—sort of.** The ice flows most rapidly in the center and most slowly at the sides—just as water would flow in a river.

It occasionally happens that a glacier that is creeping along at a rate of several feet per year will suddenly move hundreds of yards in a single day. This phenomenon of movement is called surging, and scientists believe that it is caused by the melting of water between the glacier and the ground.

790 **When a glacier travels through a valley, it changes the shape of that valley from a "V" (characteristic of something carved out by a stream) to a "U."** Many famous valleys such as the one in Yosemite National Park show the imprint of glacial forming.

791 **The point of farthest advance of a glacier is marked by a moraine.** A glacier is like a giant bulldozer, scooping rocks and debris in front of it as it goes. When a glacier stops advancing and starts retreating, a pile of debris, called a moraine, is left behind.

792 **Large glaciers have covered the Northern Hemisphere in the recent past.** From about twenty to ten thousand years ago, a large glacier covered most of North America, extending all the way down into what is now northern Illinois and Wisconsin. This is just the most recent episode among many examples of worldwide glaciation.

793 **When the glaciers grow, sea level goes down.** Since the total water budget of the earth is fixed, when more water gets taken up in ice, there is less left to fill the ocean basins. Consequently, sea level goes down during periods of glaciation. During the last glacial episode, the east coast of the United States was about 150 miles farther east than it is today.

Earthquakes

794 **Earthquakes result from the release of energy stored in rocks** —energy generated when rocks are put under compression or tension and react by deforming slightly. Eventually, the breaking point is reached and the rock snaps, releasing its stored energy. This energy creates the earthquake.

Because rocks are moving relative to each other near the boundaries

The San Andreas fault in California, generator of North America's most feared earthquakes.

of plates, earthquakes most often occur at plate boundaries. Thus, the San Andreas Fault, representing the neutral boundary between the Pacific and North American plates, is a well-known site for earthquakes. The same is true of the region in northern Turkey and the southern Soviet Union at the boundary between the Anatolian and Eurasian plates.

795 Earthquakes create seismic waves. There are many different kinds of waves, the most important being the so-called "S" and "P" waves (the letters stand for "shear" and "principle"). The P wave is a longitudinal wave, much like a sound wave. The S wave is a transverse wave in which the rock moves up and down while the wave moves forward. These two waves travel through the interior of the earth and are a major source of information about the structure of our planet.

796 One place where seismic waves are important is in the detection of small underground nuclear explosions. The basic scientific strategy is this: an explosion pushes the surrounding rock outward in all directions, and is therefore likely to create mostly P waves. An earthquake, on the other hand, tends to pull rocks sideways, and therefore will produce more S waves. The question: how small can an explosion be and still be distinguished from a natural event?

797 The severity of earthquakes is measured by the Richter scale—technically, it's called the Richter *magnitude* scale. The Richter scale is based on the amount of energy released in an earthquake and the amount of damage caused at the surface. The scale is designed so that each increase of one step corresponds to a tenfold increase in the amount of energy released. Thus, an earthquake that measures 7 on the Richter scale is 100 times more powerful than one that measures 5.

An earthquake of magnitude 2 would probably be noticed only by scientists. Magnitude 5 would not damage well-constructed buildings, but might knock down flimsy structures. On Oct. 17, 1989, an earthquake in San Francisco measured 7.1 on the Richter scale. Some geologists fear that an earthquake of magnitude 8 will occur along the San Andreas Fault at some time in the future. If this happens, it will be a major disaster. Geologists estimate that magnitude 9 is about as big as an earthquake can get—rocks simply can't store more energy than that.

Deserts

798 **Deserts are not forever.** A desert is any region that gets less than an inch of rainfall every year, while an area getting up to five inches of rainfall would be called semiarid. Large desert regions on the earth have not always been deserts in the past and will not always be deserts in the future. In general, deserts occur in regions of constant atmospheric high pressure areas or in the "rain shadow" of large mountain ranges. The Mojave Desert in California is an example of the latter type, the Sahara Desert in Africa of the former.

Both the existence of mountains and the locations of the continents change with tectonic motion. Deserts, therefore, will come and go. There is good evidence, for example, that several hundred million years ago the Sahara Desert was covered by a glacier!

799 **The primary form of erosion in deserts is through the action of the wind,** which blows soil around, and the infrequent rains which, when they come, wash soil away. Thus, there is no process that goes on in the desert that does not go on elsewhere. As far as erosion goes, deserts are not special.

800 **Deserts are more than sand dunes,** although there are areas of sand (called ergs) in many deserts. By far the most common desert topography consists of small bits of vegetation separated by patches of bare ground.

Sand dunes are built up by a process called saltation. If the slope of the dune is low, sand grains will be picked up by the wind, moved up the dune, and then dropped. This increases the pitch of the dune. If the slope is too high, on the other hand, sand will slide down the dune and lower it.

Beaches

801 **Beaches are not forever.** Tectonic motion changes the location of ocean basins, and sea level variations due to the Ice Ages are common. Thus, what is shoreline today may be underwater or far inland tomorrow. In addition to these long-term processes, there are many natural forces that work to change beaches on time scales of decades (see the following). Thus, beaches are transitory things—enjoy them while you can.

802 The "littoral conveyor belt" moves sand along beaches. When waves come into a beach, they normally approach it at an angle. After having washed up on the beach, however, they flow straight back into the water under the influence of gravity. Thus, the net motion of water (and the sand it carries) is a sawtooth pattern—"in" in the direction of the wave, "out" in the direction of the pull of gravity. Over periods of days or weeks, individual grains of sand along a beach will be moved slowly up and down the beach as each sawtooth moves them a small amount. This transport of sand has been called the "littoral conveyor belt."

803 Storms play a great role in shaping a beach. In general, stormy weather produces large waves that move sand away from the shore. Calm weather produces smaller waves that tend to move sand back. Thus, over the course of a year, the beach will actually move—out to the ocean in the winter, when storms occur, back in to the land in the summer.

804 Trying to "preserve" a beach can be a very expensive and ultimately pointless pursuit. There is tremendous political pressure to preserve beaches—to prevent erosion. This results in the construction of seawalls to block waves and jetties to stop the flow of sand. Such procedures can do little to preserve the beach. Seawalls simply prevent small waves from moving sand back on shore, although they cannot prevent the large waves from taking sand away. The net result is that the beach in front of the seawall eventually disappears. Thus, you have the paradoxical situation that someone builds a house in order to enjoy the beach, builds a seawall to preserve the house, and finds the beach—the original reason for the house—gone.

In the same way, the construction of groins keeps sand from moving down the beach and has the effect of preserving one's own beach at the cost of beggaring one's neighbors. For example, Assateague National Seashore in Virginia, a twenty-five mile stretch of natural beach and dunes, has been seriously eroded by the construction of groins to contain sand in the neighboring resort town of Ocean City, Maryland.

Rocks and Minerals

805 Rocks are not forever. Although a large rock may seem to be a very sturdy and stable thing, from the point of view of the geologist it is ephemeral. Over long periods of time, rocks are created, only to be weathered away and replaced by new rocks.

There are many ways that rocks can be broken down and eroded. Running water, abrasion by wind-blown sand or dust, leaching by chemicals, and the effects of plant growth are all examples of the weathering process.

806 When the earth was first formed, there was no soil. As rocks started to weather because of the processes outlined, small grains were chipped off. Today, of course, these grains of rock are mixed with organic material from plants and animals, as well as with various types of bacteria, to form soil.

807 The collection of sand along the beach can be thought of as an approximation to the earth's first soil—rock remains without much organic material. The next time you're at a beach, take a close look at the sand. You will see that in one handful there are many different-colored grains. Each of these grains weathered from a different kind of rock, each located in a different place in the uplands near the beach.

808 Rocks are classified according to the way that they are formed, rather than by appearance or structure. There are three general types of rock—igneous, sedimentary, and metamorphic.

809 About 75 percent of land on the continental surfaces of the earth is sedimentary rock. Most of the rest is igneous, and only a few percent is metamorphic.

810 Sedimentary rocks are formed when waterborn material settles out into the bottoms of bodies of waters. Over the course of time, layers of detritus build up and eventually are buried deep under new layers. The resulting pressure squeezes the grains of material together, and chemicals carried by underground water form a kind of glue to turn the original material into a rock. Rocks formed this way are called sedimentary rocks. Sandstone (formed, as the name suggests, from sand), limestone (formed from the accumulated skeletons of small marine organisms) and shale (formed from layers of mud and clay) are common examples of sedimentary rock.

811 You can often recognize sedimentary

rocks from your car. Have you ever driven through a road cut where looking at the rocks is like looking at the pages of a book end on? If so, you were looking at sedimentary rocks. They retain in their modern appearance the memory of their origin as successive flat layers at the bottom of a body of water.

812 **The existence of sedimentary rocks on mountaintops is evidence for plate tectonics.** You often see sedimentary rocks at elevations of ten thousand feet or more. Since these rocks had to start at the bottom of the ocean, there must have been forces operating in the earth to lift them to their present position.

813 **Igneous rocks are formed by the cooling of molten magma from the earth's interior**—from volcanoes or lava flows, for example. "Igneous" means "fire-formed," and granite, obsidian, and pumice are examples of igneous rock.

814 **Metamorphic rocks are changed from their original form by geological processes.** For example, when a rock is buried, it can be subjected to high temperatures and pressures or to chemical activity. These agents can cause the atoms in the rock to rearrange themselves or be replaced (see the following), and this changes the nature of the rock. Marble starts out as fine-grained limestone, for example, but is changed by heat and pressure to its present form.

Pop Quiz

W*hat kind of rocks make up the deep ocean floor?* Answer: The ocean floor is created by the outflow of materials at diverging boundaries. Thus, the ocean floor is made of igneous rock.

Minerals

815 **Minerals are the basic constituents in rocks.** A mineral is any inorganic solid with an ordered internal structure and a fixed atomic composition. You can think of minerals as being tinkertoy-like structures made from different atoms. There are over three thousand known minerals, each with its own name.

816 **The structure of minerals is determined primarily by the size of atoms.** For example, the cubic structure of sodium chloride (common table salt) results from the fact that the sodium ion is much smaller than the ion of chlorine, and hence fits into

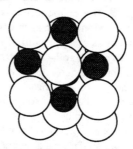

A crystal of ordinary table salt. The large white atoms are chlorine, the smaller dark atoms are sodium.

the gaps left by the packing of the chlorine ions. A true-to-scale salt crystal is shown.

817 **Crystals are a common kind of mineral** —both regular in shape and very beautiful. Like other minerals, their geometry is determined by the way that their constituent atoms are put together. The planes of symmetry of the crystal reflect the underlying arrangement of the atoms. Thus, a grain of table salt will be cubic because the atoms are arranged as shown. More complex crystal shapes can result from slight rearrangements of atoms due to differences in temperature or pressure when the crystal solidifies.

818 **The structure of a mineral can change under the influence of temperature or pressure.** If atoms are pushed together by high pressure, or subjected to high temperature, they can rearrange themselves in a mineral without changing their overall atomic composition. They will retain their new orientations even when the pressure or temperature goes back to what it was originally. Thus, minerals can serve as indicators of the past presence of high temperatures or high pressures.

819 **Grains of minerals are made from atoms, and these grains, in turn, come together to form rocks.** Often rocks are made from more than one kind of mineral. This is particularly true of sedimentary rocks, where grains of many different minerals come together to form the rock itself. Sandstone, made from the cementing together of many different kinds of grains, has this kind of structure.

The Earth's Water

820 **Water on the earth moves in cycles—** molecules of water follow a never ending path through the Earth's atmosphere, oceans and biosphere. They are evaporated from the surface and enter the atmosphere, where they may form clouds. Then they return to the surface of the earth as rain, snow, or ice. Water may be taken temporarily into living tissue or be stored in bodies of water like lakes and oceans, but sooner or later it reenters the cycle.

Scientists sometimes refer to all the water cycling through the various systems on the earth's surface as the "hydrosphere."

821 **Fresh water is a resource,** since most of the water on the earth is in the salty oceans. In the United States, we use only about 7 percent of the rainwater that falls on our land's surface. About 71 percent returns to the atmosphere through evaporation, and 22 percent is in lakes and streams, which therefore constitute an important reservoir of fresh water. Almost any water that you're likely to encounter will have come from underground or from a reservoir somewhere and will therefore have materials—both organic and inorganic—dissolved in it. When too much organic material is dissolved in water, it tends to smell and taste bad and harbor harmful bacteria.

Oceans

822 **Three-quarters of the earth's surface is covered by oceans**—indeed, from space, the planet would appear to be composed mostly of water. For most of human history, the oceans were unknown territory—regions to be crossed as expeditiously as possible. Today, they are increasingly a frontier of science.

Oceanography, the branch of science devoted to the study of oceans, combines all the sciences. It deals with energy balances and chemical reactions in oceans (physics and chemistry) as well as with living things found within the oceans (biology). The physical and biological parts of the ocean interact with each other, affecting and being affected, and neither can be ignored.

823 **Water in the oceans circulates.** The main motion of surface water in the ocean comes from large circular currents called "gyres." The water in gyres moves clockwise in the northern hemisphere and counterclockwise in the southern. The most familiar aspect of the gyres is the great ocean currents which constitute their outer margins. For example, the Gulf Stream, which runs from Florida to Europe across the north Atlantic, is one segment of the north Atlantic subtropical gyre. The Gulf Stream carries warm water from the tropics to northern Europe and is largely responsible for the relatively mild climates experienced in the latter region. The cold water returns to the tropics along the coasts of Europe and north Africa in what is called the north Atlantic current.

824 **The farther you go below the surface of the ocean, the colder and saltier it gets.** Light from the sun can only penetrate a few hundred meters below the surface of the ocean, so this is the only part of the ocean that has its own heat supply. By the time you go down to a depth of one thousand meters, the water temperature is four degrees or less. The region of transition between warm surface waters and cold bottom waters is called the "thermocline," and the upper, warmed region is called the "mixed layer."

825 **Almost all the productivity in the ocean is in the mixed layer.** Light supplies energy to run photosynthesis in plants, and plants form the basis of the food chain for animals. This means that when you get a few hundred meters below the surface of the ocean, the ability of the water to sustain life drops off drastically. This is why bodies of water like the Chesapeake Bay, which never gets more than a few hundred feet deep, are so productive.

826 **The deep ocean does not circulate very fast.** Water achieves its maximum density at four degrees centigrade. This means that the water at the bottom of the ocean will be at this temperature and water higher up will be at a slightly lower temperature. For all practical purposes, this deep ocean water forms a vast reservoir that interacts very little with the rest of the earth system.

827 **Cold water from the Antarctic ice sheet sinks and forms a slow current along the bottom of the ocean.** Almost all of the water at the very bottom of the ocean starts as ice melting in the Arctic and Antarctic and then works its way toward the equator.

828 **Ocean circulation patterns are not forever,** because the ocean basins themselves change over geologic times. For example, 50 million years ago Australia broke off from Antarctica, opening what is now called the Drake Passage around the pole. This event allowed both the oceans and the atmosphere to set up circumpolar currents in the southern hemisphere, and these currents have a profound effect on the weather at the south pole. Some scientists argue that without the currents there wouldn't even be an ice cap.

829 **To a reasonable first approximation, the deep ocean floor can be thought of as a level plain about fifteen thousand feet below sea level.** There are, of course, exceptions—deep trenches, rises, underwater mountain ranges, and so on—but if you think of the ocean this way, you won't go too far wrong.

830 **Every continent is surrounded by a continental shelf.** This is material that is actually part of the continent but which happens to be under sea level at the present time. The continental shelf typically extends out some tens (or perhaps hundreds) of miles from the seashore. Because the water over the shelf is relatively shallow, these regions of the ocean are most productive.

831 **When large amounts of sediment accumulate on the continental shelf, they occasionally "slough off" and run down the incline into the deep ocean.** When this happens, they carve out huge undersea valleys and trenches. During the flow, the water holds so much material in suspension that it becomes a sort of slurry—neither liquid nor solid. Such a flow is called a turbidity current.

832 **Sea level isn't really level**—it's not a geometrically flat plane. Where there are unusual concentrations of mass underneath the ocean floor, the force of gravity can (and does) pull the surface down. There is, for example, a "hole" about three hundred meters deep in the middle of the Indian Ocean. Of course, it's a thousand miles across so you wouldn't notice any change in altitude if you were in it.

Ocean Chemistry

833 **The sea is salty, but it's not getting saltier.** The characteristic taste of seawater comes from the fact that large amounts of sodium chloride (ordinary table salt) are dissolved in it, along with many other minerals. If you could get even a small fraction of all the gold that's dissolved in the sea out in solid form, for example, you would be a very rich person.

It used to be thought that the sea was a passive kettle into which rivers brought minerals from the land, and in which these minerals became more concentrated as water evaporated. In fact, this is not the case. Evidence from old salt beds has convinced scientists that the seas were no less salty millions of years ago than they are today.

834 **Atoms come into the sea, stay for a while, and then are removed from the water by one sort of chemical reaction or another.** For example,

when calcium is brought in from the weathering of limestone, it is taken up into skeletons of marine organisms. When these organisms die, their skeletons fall to the bottom and begin forming new limestone, which will eventually be lifted up to be weathered again. No atom stays in the sea forever.

The average amount of time an atom stays in the ocean before it is removed is called its residence time. The residence time for calcium, for example, is 850,000 years while for sodium it is 48 million years.

835 **Chlorine is said to have an infinite residence time,** but this is a little misleading. What happens is that chlorine is taken up into the atmosphere in the form of salt spray and then rained back out. Thus, although the total amount of chlorine in the ocean remains the same, a given chlorine atom doesn't necessarily spend all its time there.

836 **The ocean's salinity once fooled scientists.** Acting on the belief that the sea was becoming progressively more salty, scientists in the eighteenth and nineteenth centuries tried to estimate the age of the earth by measuring the amount of salt being carried into the ocean and calculating how long it would take for the ocean to achieve its present level of saltiness given that sort of input. The ages they got for the earth were in the neighborhood of 100 million years. Today, we know that what they were really measuring was the cycle time for materials, and not the age of the earth.

The Atmosphere, Weather, and Climate

The Atmosphere

837 The earth's atmosphere extends several hundred miles into space, although the great bulk of the gases are located within a few miles of the earth's surface. It is approximately 78 percent nitrogen by volume, and 21 percent oxygen. The bulk of the remaining gas is made up of argon (0.9 percent), carbon dioxide (0.03 percent) and varying amounts of water vapor, along with dust and other particulate matter.

The lower part of the atmosphere, the part in which we spend almost all of our time, is called the troposphere. It has an average thickness of about seven miles. Above this is a layer of much thinner air called the stratosphere, which goes to a height of about twenty miles. The temperature of the air falls steadily with altitude in the troposphere, then stays constant and finally increases as you go up through the stratosphere.

Beyond the upper reaches of the stratosphere, an increasingly tenuous collection of molecules extends out several hundred miles into space and eventually merges into the thin structure of interplanetary space. About fifty miles above the surface of the earth there is a region where light from the sun creates ions. This is the ionosphere, a region which reflects radio waves and plays an important role in long-distance communication.

838 The general circulation of the atmosphere is caused by the fact that the tropics are warmer than the poles, which gives rise to a classic convection cell. The warm air at the equator rises and migrates poleward, while cold polar air sinks and moves down toward the equator.

If the earth did not rotate, the prevailing winds at the surface in the northern hemisphere would be toward the south, as cold air masses moved from the poles to the equator.

839 Prevailing westerlies, trade winds, and all that are caused by the earth's rotation, which produces three convection cells on the planet (rather than just one)—one in the tropics, a second in the temperate zones, and a third at the poles. The earth's rotation "stretches" these cells so that the surface winds blow east and west. Near the tropics the winds at the surface blow toward the east. This is the region of the trade winds (so named because in the days of sailing ships, those wishing to travel to the New World would normally drop down into the tropic regions to take advantage of them). In the middle latitudes of the northern hemisphere, the prevailing winds come from the west. This is why if you want to know what the weather's going to be like on the East Coast of the United States in a few days, you can make a good guess by looking at what is happening in the Midwest. The third convection cell in the northern Arctic region also has eastward flowing winds.

If a planet rotates very fast, it can develop many of these kinds of convection cells. Jupiter, for example, has eleven.

840 The regions at the boundaries of convection cells in the earth's atmosphere have little wind blowing along the surface—all the motion is up and down. In the days of sailing ships, such areas were to be avoided. The region of stagnation near the equator is called the doldrums and ships sailing into it could be becalmed for long periods of time. The region of stagnation between the tropics and the temperate zone was called the "horse latitudes." There are many stories as to how it got this name. The story I heard (and I make no representations as to its accuracy) is that when ships sailing to the New World got stuck in these latitudes, they would jettison their cargoes of horses as they ran out of feed. Thus, a common sight in these latitudes would be the carcasses of horses.

Frankly, given the fact that these waters are also full of sharks, I would take this story with a grain or two of salt.

841 The jet stream separates warm from cold. A flow of air in the upper atmosphere in the form of a flattened tube several thousand miles in length, the jet stream is a hundred or more miles in width and about a mile thick. Roughly speaking, it marks the boundary between the arctic air mass and the warmer air of the mid-latitudes. In the northern hemisphere, the jet stream circles the pole.

Typical velocities can be hundreds of miles per hour, and disturbances of the jet stream are the cause of freak weather conditions. For example, sudden, prolonged, and severe cold spells in the North American winter are often caused by displacements of the jet stream.

842 The jet stream was discovered during World War II when military aircraft capable of speeds of three hundred miles an hour or more found that even with their throttles open, they would sometimes remain in a stationary position with respect to the ground. They had unwittingly entered the jet stream and tried to fly "upstream" against the current.

843 The general circulation of the atmosphere is responsible for long-term weather patterns on the earth, since this circulation determines cloud cover, rainfall, and temperature at the surface. I wish I could say that scientists today have a good understanding of the circu-

lation of the earth's atmosphere, but I can't. We do have computer models, called GCM (general circulation models) that can do a reasonable job of predicting large average global trends, but when it comes to describing regional effects (a Midwestern drought, for example), these models simply have not been developed to the point where we can trust them.

Weather and Climate

844 **Weather is short-term, climate is long-term**—the former refers to things like daily temperatures, humidities, and rainfall, the latter to long-term trends in these same variables.

845 **A front is the dividing line between a warm and a cold air mass.** In general, when a front passes there is a marked change in the weather. For example, an approaching cold air mass will slide under warmer air, lifting that warm air and causing the formation of clouds with possible rain or snow. The passage of a warm air mass into a region previously occupied by cold air may force the warm air up over the tail end of the cold air mass and bring on a day or more of drizzle.

846 **On the scale of a continent or less the movement of air tends to be dominated by the existence of high and low pressure areas.** Air will flow away from a high pressure

area and into a low pressure area under the influence of the pressure force. The day-to-day variation of the weather, then, depends on the movements of these areas.

847 **Air pressure is measured with a barometer.** The barometer is a partially filled tube open to air on one end and with a vacuum on the other. The height of the liquid (usually mercury) balances the column of air above the open end, and moves up and down with the air pressure. A falling barometer corresponds to a situation in which the air pressure is dropping. This usually signals a coming storm. A rising barometer means the air pressure is increasing, which means that a high pressure zone and fair weather is on the way.

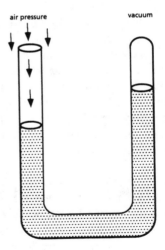

A barometer.

848 **The movement of water vapor in and out of the air is an important factor in determining weather.**

If for some reason water condenses from the air (e.g., if the air is cooled), energy will be released into the atmosphere. On the other hand, if water in liquid form evaporates, heat is removed from the air. The movement of water back and forth from liquid to vapor, then, corresponds to the movement of energy. This process plays an important role both in normal weather and in storms.

849 The local flow of wind and rainfall patterns can be affected by geographical features. One common example is the so-called "rain shadow" of a mountain. If air approaches a mountain from the west, it is forced up in order to go over the mountain's mass. The rising air is cooled and its moisture condenses out in the form of rain. On the far side of the mountain there is no more moisture in the air and consequently rainfall tends to be very sparse. Many of the arid regions in the western United States exist because they are rain shadows of the various ranges of the Rocky Mountains and the Sierra Nevada.

850 Air flows from high to low pressure regions in a curved path because of the rotation of the earth. To see why, suppose that there is a low pressure area over Miami so that air over New York City starts moving south. While that air is on its way, the rotation of the earth will carry Miami farther east than it was originally. The air moving toward the low will have to play "catch up" and follow, and it will have to continue to do so as the earth carries Miami farther and farther along. The result is that the air will follow a curved counterclockwise path as it flows.

Although the deviation from straight-line motion is due to the fact that the earth is rotating, early physicists liked to imagine that a force was causing it. They called it the Coriolis force, after the French scientist Gaspard de Coriolis (1792–1843).

Pop Quiz

I*n what direction will air flow as it enters a low pressure area in the southern hemisphere?* Answer: Clockwise.

851 Hurricanes result from the presence of low pressure areas over the ocean. Warm, moisture-laden air from the ocean surface flows into a low pressure area and is pushed upward by other air coming in behind it. As it rises, the air cools and water vapor condenses out. The energy associated with this transformation is then available to keep the hurricane going. Hurricanes actually "feed" on the warm water and can build up to quite impressive storms.

Pop Quiz

W*hy do you never get a hurricane in Kansas?* Answer: The hurricane can only travel small distances over land before it runs out of energy. Kansas is too far inland.

852 **The naming of hurricanes started with the Army Meteorological Service in World War II.** Originally, hurricanes were named after women—thus, you might get Abigail, Betty, Claudia, and so forth. In the interest of sexual equality, hurricanes are now alternately given male and female names.

The fact that we never get to hurricane Zelda (or Zeke) reflects the fact that there are seldom more than a dozen or so serious storms generated in the Atlantic each year.

853 **Hurricanes, typhoons, and monsoons are not all the same**—in fact, a storm of the type just described is called a hurricane if it occurs in the Atlantic Ocean and a typhoon if it occurs in the Pacific. Both types of storms are technically called "tropical cyclones."

The monsoon, despite the similarity in name, is not a single storm but the name for the winds that bring the rainy season to the Indian subcontinent.

854 **Tornadoes are also associated with rapid rotation of the air,** although they are much smaller in extent than a hurricane. They occur when warm and cold air masses come together. If conditions are such that the cold air is above the ground while the warm air is underneath it, violent wind motions begin as the two masses try to change positions. Typ-

A tornado.

ically, a line of thunderstorms will form and the characteristic funnel of tornadoes will come down from the bottom of dark clouds. In the United States, tornadoes are most common in the Midwest and occur most often during the spring "tornado" season.

855 Despite the differences in their appearance, all clouds are nothing more than water vapor (or in some cases, ice). The water in clouds is very thinly scattered. The typical puffy white cloud that you see on a summer's afternoon may contain no more than twenty to thirty gallons of water—barely enough to fill a bathtub, even though the cloud may be many miles across.

Clouds are formed by rising air currents. As warm air rises, it reaches an altitude where the temperature is so low that the air can no longer hold all its water. At this point, a cloud starts to form from the water droplets. The warm air continues to rise and lose moisture until it has come into equilibrium with the air around it.

Pop Quiz

Why do the bottoms of all clouds on a given day seem to be the same height above the ground? Answer: Because all the rising currents of air are cooled to the same temperature, and therefore begin dumping water, at the same altitude, which then constitutes the "floor" of the cloud layer.

856 There are many kinds of clouds, each one corresponding to different weather conditions. The puffy white clouds you see on a summer's day are called cumulus clouds (Latin for "heap" or "pile") and are created by bubbles of warm rising air. The clouds you see on an overcast day are called stratus clouds (Latin for "stretched out") and are formed from the upwelling of broad expanses of air. Feathery clouds, high in the sky, are called cirrus clouds (Latin for "fiber"). These are generally the precursors of a change in the weather and are often made entirely of ice crystals. They are typically six miles above the ground. Finally, the dark thunderhead or cumulonimbus cloud occurs during a storm itself (see diagram). Note that the typical thundercloud is very tall, with its lower face typically a mile above the ground and its top some ten miles above that. This means that thunderclouds can extend to the top of the troposphere.

Different clouds normally occur at different heights, as shown in the sketch.

857 Although people didn't begin to name clouds until the eighteenth century (presumably because they were much too common to be given serious study), we now know of thousands of different variations on the basic types shown. In fact, there is a two-volume publication called *International Clouds Atlas* that catalogues these forms in excruciating detail.

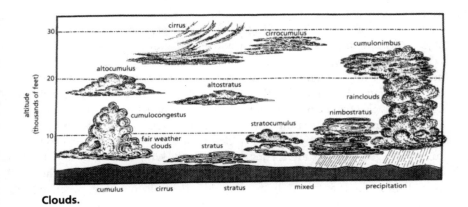

Clouds.

Enduring Mystery

858 **How do charges get separated in a thundercloud?** Ever since Benjamin Franklin first discovered the electrical nature of lightning, scientists have been trying to determine, without a great deal of success, how

the thundercloud structure comes into existence. There are two main points of view. In the so-called precipitation theories, it is noted that heavy particles of water and/or ice will fall through the cloud under the influence of gravity. The collisions between these particles and their lighter compatriots (which tend to remain suspended) is supposed to

Anatomy of a thunderstorm.

transfer electrical charge in the same way that friction does. Thus, falling objects acquire a negative charge while those higher up acquire a positive one. In convection theories, on the other hand, it is assumed that light, positively charged particles are lifted upward by convection currents in the cloud, while heavy, negatively charged particles are carried to the bottom by downdrafts. Neither of these theories can fully account for the complex structure of a real raincloud.

Most thunderstorm research centers on studies of the collisions between different-sized particles of ice within the cloud, and the charge transfers that accompany them.

859 **The separation of charges in a thundercloud produces the lightning bolt.** The large negative charge in the lower face of the clouds (ignore the smaller screening charges for the moment) repels electrons from objects in the ground underneath it. The result is a "shadow" of the cloud in the ground—a region of positive charge. These charges follow along underneath the cloud, running up and down trees and buildings as they do so.

A lightning bolt begins when the charge in the cloud gets strong enough to ionize the air in its immediate vicinity. It opens a passage of ionized air several hundred feet long, called a "leader." Because ionized air is a good conductor, the negative charge runs down into the leader. This process is repeated— another leader is formed—and the

chain of leaders makes its way almost to the ground. About 100 feet above the ground, it is met by a leader formed by the positive charges coming up to meet it. The result of all this activity is a jagged path of conducting material between the negative charge in the clouds and the positive charge on the ground. With nothing to stop it, the positive charge runs up to the cloud, neutralizing the negative charges there. We perceive this motion of charge as a lightning stroke. The energy dissipated because of the resistance in the ionized path heats the air and pushes it away. The air then returns into this partial vacuum and creates a thunderclap.

860 **The amount of charge that actually flows in a lightning bolt is not very large—** about as much as flows into your toaster in any one-second period. Since the bolt lasts only a fraction of a second, however, the power it generates is prodigious. A typical lightning bolt will generate a few hundred megawatts of power—as much as a medium-sized nuclear reactor.

861 **Lightning bolts tend to strike tall trees or buildings repeatedly.** The existence of a lightning bolt depends on the creation of an ionized path between the cloud and the ground, and it is easier to create this ionized path for short distances through the air than for long ones. The Empire State Building in New York City has been struck hundreds of times by lightning bolts.

862 The lightning rod, invented by Ben Franklin, is a metal point on a building connected by a conducting path to the ground. It ensures that when a lightning bolt strikes a building, the current passes through the metal conductors rather than through the building.

Climate

863 The earth's climate has changed radically over geological times. Over long time frames, the motions of the continents themselves are the dominant factor in determining climate. On a somewhat shorter time scale, climate can depend crucially on geological factors like the flooding of shallow seas. This phenomenon arises because water tends to absorb heat from the sun, while land tends to reflect it. Thus, 65 million years ago, when the western United States contained a vast inland sea, the climate of North America was much warmer than it is today.

864 Solar variability also plays a role in determining climate. Scientists speculate that changes in solar brightness—perhaps less than 1 percent—occur regularly. Even changes of this magnitude could be expected to affect the earth's climate, although we cannot predict how at the moment.

865 Several times during its history, the earth seems to have been subjected to a rapid succession of ice ages— indeed, we live in one of those periods right now. The current theory is that ice ages are caused by small (but regular) changes in the shape of the earth's orbit and the direction of the earth's axis. When these small effects reinforce each other, more snow accumulates during the winter than melts during the summer. The result: more snow stays on the ground in the summer, and this snow reflects more heat, which in turn causes more snow to remain for the following summer, and so forth. When this happens, glaciers move out from the poles and high mountain ranges to cover large parts of the continents. This process is called the Milankovitch cycle, after the Serbian engineer who first understood it.

866 El Niño events occur regularly, every two to seven years, and produce unusual weather. The name comes from the Spanish for "Christ Child" because the weather starts to change around Christmas. An El Niño event begins with a warming of the waters off the western coast of South America and is followed by unusual weather throughout the western hemisphere. In 1982–83, for example, there was serious flooding in South America and a series of major storms along the California coast. Some scientists attribute the severe drought of 1988 to El Niño.

The current theory of El Niño is that it is sustained by the water in the Pacific Ocean, which sloshes around like water in your bathtub. When the warm surface water sloshes toward South America, wind patterns are set up that drive water to the west, thereby keeping the warm water in place for a time. When the water sloshes back, cold water comes up near the coast, setting up other winds that are characteristic of normal weather patterns. This cycle has been repeating for a long time, and appears to be a relatively permanent feature of the earth's climate.

Climate Issues

867 The amount of carbon dioxide and other so-called "greenhouse gases" in the atmosphere is another determiner of climate. These gases are transparent to visible light but absorb the infrared. Thus, energy that the earth would normally radiate back into space remains trapped in the atmosphere, warming it. It is believed that the planet Venus is as hot as it is because of the greenhouse effect.

868 The greenhouse effect is a major policy issue. Every time you drive your car or burn gas in your stove, you are adding carbon dioxide to the atmosphere. There is a great worry that human beings, by burning fossil fuels, are in the process of provoking a sudden warming in the earth's atmosphere. Whether this warming has already started is still a matter of debate, but it seems reasonable to suppose that if you keep adding carbon dioxide and other gases to the atmosphere, the greenhouse effect will eventually take over. At present, the estimates are that greenhouse warming might involve a change in the average temperature of several degrees centigrade, almost as much as that which followed the last period of glaciation. Finding ways to mitigate this effect is a high-priority item for policymakers. There is little we can do to change the thermodynamics of the atmosphere, so efforts are concentrated on reducing carbon dioxide emissions and preventing the destruction of forests, whose trees remove carbon dioxide from the atmosphere as they incorporate the carbon into their tissues.

869 High in the stratosphere there is a thin layer of ozone, a molecule made of three atoms of oxygen. The ozone layer is primarily responsible for absorbing ultraviolet radiation from the sun, so its existence is extremely important for life on land. Recently, the emissions of chemicals known as chlorofluorocarbons (CFC) have produced catalytic reactions in the stratosphere which have the effect of thinning out the ozone layer.

In 1984, scientists noticed a serious thinning of the ozone layer over Antarctica during the southern hemisphere spring months. This "ozone hole" is caused by a combination of events peculiar to the south polar region, having to do with reactions taking place on ice crystals in clouds that form during the months of darkness.

In 1986, an international convention held in Montreal adopted a protocol calling for the reduction in the use of CFCs by 50 percent and further reviews of the ozone problem in 1990. A total elimination of CFCs will probably be adopted at some time in the near future. The ozone problem, unlike the greenhouse effect, can be solved at relatively little cost.

870 When coal is burned, compounds of sulfur and nitrogen are emitted from smokestacks along with the carbon dioxide. Similar wastes come from automobiles. In the air, the sulfur and nitrogen compounds undergo chemical reactions to form nitric and sulfuric acids, which then are washed out by the rain. This so-called acid rain is held to be responsible for a number of harmful environmental effects. Among them are the destruction of forests in the northeastern United States and Canada (although the role of acid rain in these regions has been disputed), and western Europe, as well as the destruction of the fabric of buildings and monuments throughout the world.

Methods of combating acid rain include "scrubbing" smoke from large coal-fired furnaces to remove the harmful compounds and limiting emissions from cars.

7

ASTRONOMY

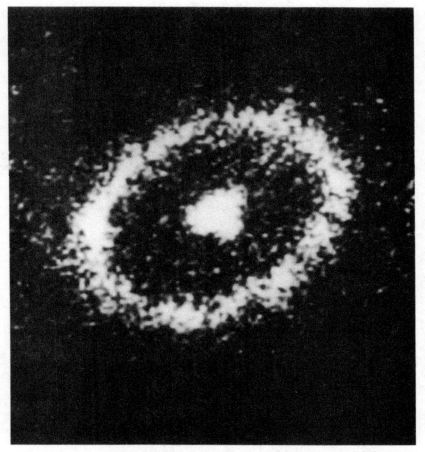

The supernova of 1987 (see item 885). The bright ring is material that was thrown out of the star and is being returned to space. This sort of process is the source of all heavy elements on earth.

Stars

871 **Stars, like everything else, are born, live out their lives, and die.** It was only fairly recently in the history of the human race—the nineteenth century, to be exact—that people realized that stars couldn't last forever. Stars are continually pouring energy into space, and that energy has to come from somewhere. Today, we know that the sun, like most stars, burns hydrogen to produce that energy. But even for a huge body like the sun, the supply is not endless. The sun, like a campfire at the end of an evening, will someday stop burning and die.

There were some interesting attempts to explain the energy output of the sun. In the nineteenth century scientists showed that if it were made of pure anthracite coal (the best fuel known at the time) it could only last for 10,000 years at its present rate of energy output.

872 **The energy source of stars is nuclear fusion.** Deep inside the sun, nuclei of hydrogen come together in a series of reactions whose end product is helium and some excess energy. The sun consumes hydrogen at the rate of 700 million *tons* per second, and has done so since shortly after it formed. Most other stars generate energy in the same way for most of their lifetimes, only going on to other things when the hydrogen is exhausted.

873 **The energy in the sunlight that came in your window today started out in the core of the sun 30,000 years ago** —just after the last Neanderthal passed from the scene. It spent most of its lifetime slowly making its way through the jostling crowd of atoms in the sun, then had a brief eight-minute sprint through empty space to come to the earth.

874 **A star's life is a battle between the nuclear fires and gravity.** The force of gravity is always pulling the star in on itself. For a while—as long as the fuel lasts—the star can maintain a precarious equilibrium by using the energy from nuclear reactions to balance the inward pull. The life of every star is a battle between these two competing forces. Eventually, the fuel must run out and gravity will win. It is the victory of gravity that we refer to as the death of the star.

875 **Not all stars are like the sun.** If you think of the sun as being roughly the size of a basketball, the range of other stars would go from those the size of a grain of sand to those the size of a large building. Stars come in all brightnesses, colors, and many very exotic forms.

Amidst all this variety, the sun is a very ordinary star. It is average in its lifetime, its chemical composition, and its luminosity. There is absolutely nothing to distinguish it from its brethren in the Milky Way.

The H-R Diagram

The Hertzsprung-Russell (H-R) diagram, developed in 1905 by the American astronomer Henry Russell and the Norwegian astronomer Ejnar Hertzsprung, gives us a good way to visualize the tremendous variety of stars. The idea is to make a graph whose vertical axis is the brightness of the star and whose horizontal axis is the star's color or temperature. Every star appears as a single dot on this graph—the approximate position of the sun, for example, is indicated by the arrow. Most stars fall on a line running from upper left to lower right. This is called the main sequence, and stars in it (like the sun) are called main-sequence stars. The stars in the upper right-hand corner of the H-R diagram are cool but give out a great deal of light. These are the so-called "red giants." The stars in the lower left-hand corner of the H-

R diagram are dim but hot. These are called "white dwarves" (see the following).

876 **The sun is middle-aged.** It began burning up hydrogen about 4.6 billion years ago and has now gone about halfway through its expected lifetime. In its life expectancy, as in most other things, the sun is a typical star.

877 **Stars tend to be grouped together in the sky.** About two-thirds of all the stars you see are in double-star systems—that is, systems of two stars in orbit around each other. In addition, the galaxy has many large clusters of stars, ranging in size from hundreds to millions of stars each.

878 **The brightness of a star is measured in terms of its "magnitude."** Before the invention of the telescope, stars were grouped by what we would today call their apparent magnitude—that is, their brightness as seen from the earth. The brightest stars were said to be first magnitude,

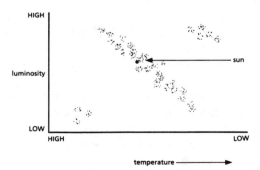

The Hertzsprung-Russell diagram. Notice that the temperature decreases as you move to the right along the horizontal axis.

the next brightest second magnitude, and the dimmest that can be seen with the naked eye sixth magnitude. This scheme was retained by astronomers even after the invention of the telescope.

Each drop in magnitude corresponds to a drop of 2.5 in the brightness of the source as seen from earth. Thus, a sixth magnitude star is approximately 100 times dimmer than a first magnitude. It is not at all unusual today for astronomers using state-of-the-art telescopes to detect twenty-fourth magnitude objects in the sky.

The apparent brightness of a star depends on how far away it is and on how much energy it is giving off (its "luminosity"). To eliminate the ambiguity associated with the distance to the star, astronomers have defined the "absolute magnitude" of a star as the brightness it would have if it were seen from a distance of thirty-three light-years. The absolute magnitude does not depend on the distance to a star, but measures something intrinsic to the star itself.

The Life History of a Star

879 **Stars are created out of clouds of interstellar dust.** Gravity will cause such a cloud to collapse on itself. As the cloud of condensing dust contracts, it heats up. Eventually, the temperature of the center gets to the point where fusion reactions ignite and the star is born.

Astrophysicists still argue about the fine points of the birth of stars. There appears, for example, to be a kind of "stuttering" in which the newborn star misfires like a car on a cold morning and blows large amounts of material back into space. Eventually, however, the star settles down into the steady equilibrium that characterizes the star in its mature life.

880 **Big stars live fast, die young, and make spectacular corpses.** It may seem pardoxical that big stars, with all that extra fuel, actually live shorter lives than their smaller contemporaries. The reason is fairly simple. The bigger the star, the stronger the pull of gravity trying to make it collapse (see the foregoing). The stronger the pull of gravity, the more fuel has to be burned in the nuclear fires to keep the star stable. The net result is that a star ten times as big as the sun will live only 20 or 30 million years, while a star much smaller than the sun might live as much as 100 billion years.

881 **When the sun runs out of hydrogen, it will begin to burn the ashes of its nuclear fire.** The primary fuel of stars is hydrogen. The "ash" of the nuclear burning of hydrogen is helium. When the sun begins to run out of hydrogen 5 billion years from now, it will no longer be able to balance the force of gravity. It will start to contract, and the contraction will heat up its center. The higher temperatures will, in turn, ignite nuclear reactions that burn helium. In this way, each successive nuclear burning consumes the ashes of the previous fire.

882 The sun will become first a red giant, and then a white dwarf. Near the end of its life, the outer part of the sun will start to expand. At its greatest extent it will swallow up Mars and Venus and will fill half the sky as seen from the earth. At this point, all life on earth will vanish. At this stage of its life, the sun will be a type of star called a red giant. Eventually, the sun starts to cool and contract again. This time, the temperature never gets high enough to ignite the next fire. The end product of the contraction is a star a few thousand miles across, known as a white dwarf. The force of gravity tries to make the star contract further, but cannot push the electrons in the star closer together. An eternal equilibrium is reached.

883 Large stars die as supernovae. When a large star finishes burning up its hydrogen and helium, it continues to contract and get hotter. The temperature burns up the helium, then carbon, then silicon, and eventually produces iron. Iron is the ultimate nuclear ash. You cannot get energy from iron by breaking it apart, and you cannot get energy from iron by allowing it to fuse with other nuclei. It simply won't burn. In a large star, then, the iron ashes begin to clog the core.

When the nuclear reactions stop inside a large star, the core collapses under the influence of gravity. The outer parts of the star have the "rug pulled out from under them" and start to fall inward. On the way, they meet the core, which is rebounding, and all hell breaks loose. The result is an explosion in which the star literally tears itself apart as it pours energy into space. For short times, supernovae can put out more energy than an entire galaxy.

884 The way a star dies depends on its mass. The only thing that matters in determining what the last stages in the life of a star will be is how big it is. If it's the size of the sun, or even up to five times as large, it will go through the red giant-white dwarf sequence we've described for the sun. If it's bigger than eight times the mass of the sun, it will become a supernova. If it's between five and eight solar masses, we don't know enough to predict what will happen to it, although we know it will go one route or the other.

885 Supernova 1987A was the most recent supernova in our neighborhood. Supernovae are not rare—they happen several times a century in most galaxies. In February 1987, a supernova went off in the Magellanic Cloud, near the Milky Way. It was the first supernova close enough so that it could be observed with all the techniques of modern astronomy.

The big news about 1987A (as it is known) is that there was no news.

It behaved more or less as the theories predicted. This was a great triumph for modern astrophysics.

886 A neutron star is one possible outcome of a supernova. During the supernova collapse, the electrons in the core are forced inside the protons. This reaction turns every proton in the star into a neutron. The result is a neutron star—a star about ten miles across but about as massive as the sun. The neutron star is stable because the force of gravity is not strong enough to force the neutrons any closer together than they already are. We believe that we have evidence for the existence of many neutron stars in the sky.

887 A pulsar is a rotating neutron star. There may be hot spots on the surface of neutron stars that emit radio waves. If the star is rotating rapidly, the radio waves sweep through space much as the beam from a lighthouse sweeps over the ocean. On earth, we detect these radio waves as pulses, one pulse for each time the beam sweeps by. Because we see the radiation from these stars as pulses, we call them pulsars.

888 The regularly spaced signals from a pulsar look a lot like an attempt at communication from extraterrestrials. In fact, when astronomers in England first detected them in the late 1960s, they were referred to as LGM ("little green men") signals by people in the observatory.

889 Some pulsars may be cannibals. There are roughly 500 pulsars in the sky, of which about a dozen rotate at almost unbelievable rates near 1000 times a second. Astrophysicists believe that these fast pulsars were once ordinary pulsars, rotating 300 times a second, that happened to be part of a double-star system. The intense gravitational field of the neutron star pulled mass off the normal partner. This cannibalized mass then spiraled in to the surface of the pulsar, speeding up the rotation to its present rate.

890 A black hole is another possible end state of a supernova. If the mass of a collapsing supernova core is big enough, gravity can force the neutrons together and the star evolves into a black hole—an object so massive and so small that nothing—not even light—can escape from its surface. A black hole as massive as the sun would be only a mile or so across. The black hole represents the ultimate triumph of the force of gravity over the matter in the star.

891 There is no hard evidence that any black holes exist in the universe. This surprises most people, as black holes have a prominant place in both science writing and science fiction. They are very hard to detect, however, because by definition nothing can ever get from them to us. The only way you can know they're there is by their gravitational effects, and that means we have to find a double-star system where one of the members is a black hole. Astronomers have half a dozen candidates for systems that might contain a black hole, but it is wise to remember the attitude of MIT physicist Philip Morrison, who said about evidence for black holes, "I'll believe it when I see it."

892 A nova (as opposed to a supernova) refers to any star which appears suddenly to become bright in the sky. What we now call a nova is in fact a double star system in which one of the members is a white dwarf. Mass from the large star falls on the surface of the white dwarf until it accumulates to a depth of a couple of feet. Then, because of the tremendous pressure and heat, the extra mass ignites in nuclear fires and burns itself off. We see this ignition as a brightening of the star in the sky. The same nova can go off many times, and the typical time between brightenings is about 10,000 years.

893 The starry sky is simply a transitory phase in the evolution of the universe. White dwarves do not have nuclear fires at their centers, but still are bright in the sky because they radiate stored heat. As this heat leaks away into space, the white dwarf will eventually stop shining. It will become a brown dwarf or a black dwarf—a cinder in the sky. Similarly, pulsars will eventually radiate all their energy into space, stop spinning, and become another kind of cinder. Eventually, there will be no stars left in the sky.

Stars and Chemistry

894 Stars are the factories in which heavy elements are made. The big bang produced mainly hydrogen and helium, the raw fuel of stars. Fusion reactions in stars produce all other chemical elements. In a very real sense, then, the stars are the cauldrons in which the stuff of the universe is forged.

If the first stars formed were like those we see today, some early stars must have been large. These stars would have burned themselves out quickly, fabricating the nuclei of heavy elements as they did so. When these stars died, they became supernovae and the elements were returned to the interstellar medium, where they became the starting point for the formation of second- and third-generation stars. Thus, as the galaxy aged, the complement of heavy elements increased. The sun and the solar system, having formed fairly late in the galaxy, incorporated these star-made elements into their structure.

895 Almost all the heavy elements in your body were made in supernovae somewhere. All elements heavier than iron, and most of the atoms of elements heavier than helium, are made in supernovae and then returned to the interstar medium when the supernova explodes.

There they wait until they are taken up in the formation of a new star and (perhaps) planets. The sun and the earth were formed from this sort of enriched gas 4.6 billion years ago.

The calcium in your bones, the iron in your blood, and the carbon in your tissues all got their start inside a star somewhere, and most likely inside of a supernova.

Galaxies

896 **There are no rogue stars in the universe.** When we look into the sky, we see stars grouped into large collections called galaxies. It doesn't have to be that way. The stars could have been spread out uniformly, they could all be clustered into one gigantic galaxy, or you could have had almost any kind of distribution in between. So why are they arranged in bunches instead of some other way?

That's a question that scientists really can't answer yet.

897 **The sun is part of the Milky Way galaxy.** Ours is a very ordinary galaxy. It has about 10 billion stars, and its most obvious feature is that its bright stars are in spiral arms. From a distance, our galaxy would look like a flat, pancake—shaped disk about 80,000 light-years across,

An artist's view of the Milky Way as it would be seen by an astronomer in another galaxy.

with four spiral arms running through the disk. At the center is a large spherical concentration of stars called the nucleus. Our sun is located about ⅔ of the way out along one of those spiral arms.

898 Stars in the central nucleus of a galaxy are tightly packed. Out near the sun, stars are located many light-years apart. In the center of the galaxy the distance between stars becomes much smaller—perhaps only a few times the size of the solar system. If you were on a planet in orbit around one of those stars, there would be no night. Even when your side of the planet was facing away from your particular sun, there would be plenty of light from the other stars in the neighborhood.

899 The Milky Way, visible in the summer sky, is the result of looking through the disk of the galaxy. The name for our galaxy comes from the Milky Way formation, a bright band of thousands of visible stars stretching across the sky. If you imagine the galaxy as a flat pancake with the sun inside, then the Milky Way is what you see when you look out into the dough.

Pop Quiz

Why don't you see a lot of stars outside the Milky Way? Looking away from the Milky Way corresponds to looking out of the pancake, rather than through it.

900 The word galaxy comes from the Greek word for milk, *galacticos*. The sight of the Milky Way must have reminded the Greeks of milk being spilled across the sky.

901 The Milky Way, like all galaxies, has a great deal of structure. If you approach a galaxy like the Milky Way from a great distance, the first thing you would encounter are small "suburban" galaxies—things like the Magellanic Cloud. Moving closer to the galaxy you would encounter a spherical scattering of globular star clusters, each with a hundred thousand to a million individual stars. Coming inward from the clusters you would see the familiar pancake and spiral arms that most of us associate with galaxies, and, finally, the central cluster of stars in the galactic nucleus. Add the fact that this whole complex structure is enclosed in an invisible sphere of dark matter (see the following), and you can see that even though a galaxy may seem like a simple thing, it isn't.

It is likely that there is a black hole at the center of the galaxy. Astronomers studying radiation coming from the center of our galaxy (in the constellation Sagittarius) have come to the conclusion that something very strange is going on there. They see a large empty space in the center, free of gas but surrounded by swirling, chaotic threads of material. From the motion of this material, they conclude that there must be a massive object at the center of the galaxy—several million times bigger than the sun. The best candidate for such an object would be a large black hole.

902 **The spiral arms that we normally associated with a galaxy are actually only a very small part of its total structure.** We believe that at least 90 percent (and perhaps more) of the mass of a galaxy like the Milky Way is in the form of dark matter. This dark matter occupies a sphere which completely encloses the spiral arms of the galaxy and extends far out beyond them. In other words, when you look at a galaxy what you see is most emphatically *not* what you get.

903 **We know about dark matter because we can see its gravitational effects, even if we can't see it directly.** In the case of galaxies, there are lone hydrogen atoms floating out beyond the spiral arms, revolving around the visible galaxy like a microscopic satellite. We can detect radio waves from these atoms, and from these waves we can see that their orbits are shaped by gravitational forces beyond those exerted by visible matter. The source of this extra force is what we call dark matter.

Galaxies

904 **The existence of other galaxies was not confirmed until the 1920s.**

Galaxies are such an important part of our picture of the universe that it's hard for us to realize that not too long ago there was a major debate in the sciences as to whether other galaxies existed at all. The argument was over whether cloudy patches of light in the sky were other "island universes" like the Milky Way or simply clouds of gas in the Milky Way itself. The question was finally resolved by the American astronomer Edwin Hubble on the hundred-inch telescope at Mt. Wilson in California. With this telescope he was able to see individual stars in the Andromeda galaxy, our nearest neighbor, and was able to prove that it was over 2 million light-years away.

905 **The famous German philosopher Immanuel Kant first speculated on the idea that there might be other galaxies in the universe.** He first used the phrase "island universe" to refer to them.

906 **Most galaxies are spirals, like the Milky Way**—about three-quarters of them, in fact. Spiral galaxies are flat, more or less like pancakes, and have two or four (occasionally more) curved spiral arms. Pictures of other spiral galaxies look more or less like the blade of a buzz saw.

There are other kinds of galaxies besides spirals. Of the galaxies that aren't spiral in nature, most are what are called elliptical galaxies. As the name suggests, these are large elliptically shaped collections of stars that have no other particular structure.

As you might expect with any classification scheme, some galaxies are neither spiral nor elliptical but "miscellaneous." These include things that are called "dwarf" and "irregular" galaxies. I think of them as being analogous to the bits and pieces of cookie dough you have left over from a sheet after you've applied the cookie cutters as efficiently as you can.

907 Galaxies formed by condensation from gas clouds—by a process similar to that which formed the sun and the solar system. In a large gas cloud, there were some areas where more mass collected (by chance) than in others. These high-density areas attracted neighboring matter to them, making them still more massive and therefore capable of attracting still more matter. Eventually this process would have caused a large cloud to break up into separate galaxies, and inside each galaxy the process would have continued to operate to form separate stars.

As the material in the galaxy was pulled inward by the force of gravity, whatever rotation the galaxy had to begin with was magnified. This effect is similar to what you see when an ice skater goes into a spin. If she pulls her arms in, she spins very fast. When her arms are put back out, she slows down. In the same way, the galaxy, as it condenses and contracts, "pulls in its arms" and speeds up its spin. Today, the Milky Way galaxy turns around its axis once every 250 million years.

The rotation of galaxies also explains their general "pancake" structure—the rotation throws the material from which stars are made to the outside, much as a potter's wheel does to clay.

908 The spiral arms in a galaxy are not what they appear to be. It is tempting to think that the spiral arms in a galaxy arise because the galaxy rotates, and are in some way analogous to the kinds of patterns you see in cream when you stir your coffee. This can't be true. The Milky Way, for example, has spun around many times since it was born. If the spiral arms were analogous to the cream in your coffee, they would have been "wrapped up" long ago.

Current thinking is that the spiral arms do not really represent places where there are more stars in galaxies, but simply places where the stars are brighter (i.e., younger). Looking at a galaxy is a little like flying over a town at night. Main Street may be very bright, but that doesn't mean that that's where most of the people are.

Enduring Mystery

909 Why do galaxies have spiral arms? It is thought that pressure waves travel around the galaxy more or less like spokes on a wheel, triggering the formation of stars as they go. Therefore, the bright spiral arms of gal-

axies should be thought of simply as markers of places where stars are being formed.

As striking a feature as spiral arms are, we really do not understand what it is that causes the pressure waves that give rise to them. Where does the energy to keep them going come from?

Radio Galaxies

910 **Radio galaxies are sites of galactic violence.** Galaxies like the Milky Way tend to emit most of their radiation in the form of visible light, much of our sun does. There are, however, a number of galaxies that emit very strong radio signals. These are known as radio galaxies. Galaxies that appear bright to the eye (i.e., galaxies that emit a lot of visible light) tend to be faint in the radio sky and vice versa.

When we look at radio galaxies with ordinary telescopes, we tend to see galaxies in which there is a great deal of turmoil—explosions and other kinds of behavior we do not associate with relatively tranquil places like the Milky Way. In fact, there appear to be two kinds of galaxies in the universe—violent galaxies, like the radio galaxies, and sedate, homey, hearthlike places like our Milky Way.

The violence in some radio galaxies is so great that when you look at them you can see huge jets of material being spewed out from the center of the galaxy. These jets are often many times larger than the en-tire galaxy itself—they are the most striking feature of the radio sky.

911 **Quasars are examples of radio galaxies.** Quasar is an acronym for "quasi stellar radio source." It refers to the fact that, while these objects emit copious amounts of radio waves, when people first looked at them with optical telescopes they used to show up as single points of light—like stars. We now know that quasars are very distant galaxies of the violent radio-emitting type. There are over a thousand quasars known in the sky.

912 **The most distant (and oldest) objects known are quasars.** Astronomers measure the distance to quasars by using the redshift. The most distant quasar, known as 0051-229, is approximately 16 billion light years away from the earth and almost at the edge of the observable universe.

Because quasars are so far away, the light that reaches us from them has been traveling for a long time. Consequently, the quasar we see when we look in the sky may have no relation whatsoever to the object that exists at that point today. Some astronomers believe that quasars mark an early, violent evolutionary point in the development of all galaxies. If this is so, then if you were an astronomer standing on a planet in what we call "quasar 0051-229" looking toward our Milky Way, you might see us as a quasar and see yourself as a perfectly normal, run-of-the-mill galaxy.

913 Why are there galaxies at all?

Even though we have known about other galaxies for over half a century, we still cannot explain why they exist. In fact, most theories seem to predict that they shouldn't.

The basic problem is this: galaxies cannot start to come together under the influence of gravity until the universe is about five hundred thousand years old. Before this the pressure in the expanding big bang is just too great. On the other hand, in only a short time after the five-hundred-thousand-year-mark the expansion will have spread the matter out too thinly for galaxies of the size we see to form. No one has yet figured out how to squeeze the lengthy process of galaxy formation into this tiny window of time. The effort goes on, of course, but this remains the major unsolved problem in modern cosmology.

Cosmology

914 The universe is expanding.

The galaxies, in general, are moving away from each other. This fact was discovered in 1923 by the American astronomer Edwin Hubble. The measurement he made was this: He looked at light emitted by distant galaxies and compared it to light emitted in the same atoms in laboratories here on earth. He found that the wavelength of the light from distant gal-

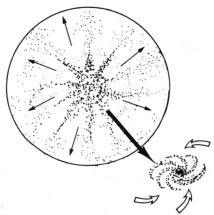

A visualization of the overall expansion of the universe. Galaxies form from local coagulations within the overall expansion.

axies was longer than expected—that it was shifted toward the red end of the spectrum. Interpreting this as the Doppler effect, he concluded that all the galaxies in the universe are moving away from us, and that the farther away a galaxy is, the faster it is receding.

915 **The expansion implies that the universe had a beginning in time.** If you imagine taking a film of the universal expansion and "running it backwards," you would see that as you move backwards in time the universe gets smaller and smaller. Eventually, the universe would be shrunk down to a single geometrical point. Clearly, this represents a beginning of some sort, and the time defined in this way is usually called the "Hubble age" of the universe. The best guess is that Hubble age lies somewhere between 10 and 20 billion years.

916 **The picture of the universe in which everything starts off in a hot, dense state and then expands is called the Big Bang.** The term is used to describe both the general evolution of the unverise and the event that started things off.

The best way to picture the Big Bang is to think about a piece of bread rising and expanding. If there were raisins scattered through the dough, each of them would be analogous to a galaxy. Standing on a raisin, you would see yourself as stationary and see other raisins moving away from you because of the expansion of the dough. The farther away a raisin was, the faster it would be moving, simply because there would be more dough between you and it. This, of course, is exactly analogous to what Hubble saw when he looked out into the universe.

917 **The term "Big Bang" started as a putdown.** In the 1940s, there were many competing theories about the nature of the universe. British astrophysicist Fred Hoyle coined the term "Big Bang" as a snide putdown of his competitors, only to have the term find its way into the general consciousness as the description of *the* correct theory.

918 **The Big Bang is not like an explosion.** It is tempting to think of the big bang as being somehow analogous to the explosion of an artillery shell. As the analogy of the raisins and the bread dough shows, however, this isn't what is happening. It is the fabric of space itself that is expanding, and the galaxies are just carried along. This may seem like a fine point, but unless you grasp it you can tie yourself in all kinds of knots thinking about the Big Bang.

919 **Galaxies do not expand, at least not very much.** Although the distance from one galaxy to the next is increasing, the size of individual galaxies stays pretty much fixed. In our analogy, the raisins don't expand—the expansion is strictly a property of the dough.

> **"What's the universe expanding into?"** If I had to draw up a list of my least favorite questions, that one would be right at the top. I don't dislike the question because it's silly—in fact, it's very profound. The reason I don't like it is that I can't come up with a satisfying answer.
>
> The act of explaining a scientific idea in everyday language is, in essence, an act of translating from one language to another. The language of science is mathematics and the language of ordinary discourse is English. Under normal circumstances, when I'm asked a question I translate it from English into mathematics, find the answer in the mathematical language, and then translate that answer back into English. The problem with this question is that it can't be translated into mathematics. It is like the question "What's north of the North Pole?" If you think about it for a while, you'll realize the problem is not that there's "nothing" north of the North Pole, but that there's not *even* nothing north of the North Pole. The frustration you're feeling with that answer is very similar to the frustration I feel when I'm asked what the universe is expanding into.

The Evolution of the Universe

920 **When the universe was younger it was hotter.** Materials are generally hotter when they're compressed than when they're not. The universe is no exception to this rule. When it was younger, it was hotter and therefore the collisions between the constituent parts were more violent. The farther back you go, the higher the temperature of the universe and the more violent the collisions. This notion is the key to understanding the evolution of the universe.

921 **The universe evolved through a series of "freezings."** If steam at very high temperature and pressure is released suddenly, it will expand, cooling as it does so. When it reaches 2128 F, an important change occurs—the steam condenses into water. The system will continue to expand and cool until another crucial stage is reached, the stage at which the water freezes into ice. Thus, we can characterize the evolution of the steam as a uniform expansion punctuated by sudden changes in the state of matter.

In the same way, the evolution of the universe can be characterized as periods of uniform expansion and cooling, punctuated by short periods when fundamental changes occur. I shall, in fact, identify six of these sudden events which I like to call "freezings."

922 **The formation of atoms, about five hundred years after the Big Bang, was the most recent freezing.** Before this time, if an electron tried to hook onto a nucleus, subsequent collisions would be sufficiently energetic to knock it off. Thus, up to the five-hundred-thousand-year-mark, matter existed in the form of plasma, and only later did it exist in the form of atoms.

Light and other electromagnetic

radiation cannot travel very far in a plasma without interacting with matter. Thus, the universe before the formation of atoms was opaque. If one part of the universe had a clump of matter in it, it would be more opaque and the ambient radiation inside the plasma would interact more strongly with it, blowing the clump of matter apart.

After atoms formed, the universe suddenly became transparent and the light was released. From this time on, light exerted very little force on matter. This means that galaxies could not have started to form until after the universe "froze" to create atoms.

923 **Nuclei formed when the universe was about three minutes old.** At this time the temperature of the universe had dropped to the point that atomic nuclei could exist. Before this time, if a proton and neutron tried to come together to form an elementary nucleus, they would be knocked apart in subsequent collisions. After this time, nuclei of light atoms could stay together. Thus, three minutes (actually three minutes and forty-five seconds) marks another important "freezing" in the history of the universe.

924 **When they formed nuclei in the early universe, the particles were working against time.** Nuclei couldn't start forming until the temperature had dropped, but if they waited too long the Hubble expansion would have spread the material so thinly that not enough collisions could occur to form significant numbers of them. Because of these twin effects, the "window of opportunity" for the building of nuclei is quite short. In fact, only isotopes of hydrogen and helium, together with very small amounts of lithium-7 (whose nucleus has three protons and four neutrons) were made in the Big Bang. All other elements in the universe were made later.

925 **At ten microseconds the quarks froze out.** From three minutes backwards to a time of about ten microseconds, matter existed primarily in the form of elementary particles. At ten microseconds, the temperature of the universe was at the point where quarks could "freeze" into elementary particles. Before this time, matter existed in the form of quarks and leptons; after this time it existed in the more familiar form of elementary particles (electrons, neutrons, protons, and others).

926 **10^{-10} seconds marked the first freezing of a force.** From now on back to the Big Bang, the freezings involve forces instead of matter. Ten billionths of a second after the big bang (10^{-10} seconds) marks the point at which the weak force unifies with the electromagnetic force. Before this time, there were only three forces in the universe—the strong, gravitational, and the unified force that physicists call the electroweak. After this time, the universe had the full complement of four forces.

927 We can reproduce the conditions of the universe as it was 10^{-10} seconds after the big bang in our laboratories. At half a dozen laboratories around the world there are machines that accelerate protons or electrons to almost the speed of light. These accelerated particles then collide head-on with each other. The result of these collisions is that for a fleeting fraction of a second a volume of space somewhat smaller than the size of a proton is raised to the temperature that characterized the entire universe when it was 10^{-10} seconds old. Thus, although our discussion of the Big Bang seems to be taking us back to times more appropriate for fairy tales than for sober physics, we can speak with some confidence about what happened during this particular freezing because we can reproduce the conditions under which it occurred in our laboratories. This statement cannot be made for the theories that follow.

928 The grand unified theories describe the universe as it was 10^{-33} seconds after the Big Bang. At that point in time, the theories tell us, the strong force unified with the electroweak. Before this time there were two forces operating in the universe; after this time there were three. The theories that characterize this transition are the so-called grand unified theories or GUTS. We cannot at the present time duplicate the conditions that obtained during this freezing in our laboratories, so from this point backwards we have to be guided more by theory than by experiment.

929 The absence of antimatter in the universe is explained by the GUTS. The way it works is this: The grand unified theories are used to describe reactions that can be done at the relatively low energies available in our laboratories. Comparing the theoretical descriptions and experimental results allows us to fix some of the numbers that appear in the theories. We then apply the theories with the experimentally determined parameters to the early universe. They predict that for every 100 billion antiprotons produced in the early universe, there were 100 billion and 1 protons produced as well. As time went on, the protons and antiprotons found each other and annihilated. All of the solid matter of the universe, including your body, is made from that little bit of left-over matter that couldn't find a way to annihilate.

930 Inflation is a feature of the GUT. This term is used to describe a period during which the expansion was very rapid. Just as the water expands when it freezes into ice, the universe expanded when the strong force "froze"—but the universe expanded much more than the water left out in a bottle overnight. Estimates are that this expansion is a factor of 10^{50}—enough to take the universe from something smaller than the smallest particle up to something the size of a grapefruit. If you expanded to 10^{50} times your present size, you would be bigger than the entire universe!

931 At 10⁻⁴³ seconds the ultimate unification occurred. The initial freezing of the universe, as you probably expect, involves the unification of the gravitational force with all the other forces. This occurred at 10^{-43} seconds, the so-called Planck time (named after the German physicist Max Planck, one of the founders of quantum mechanics). Before this time, the universe was as simple as it could be—there was only one kind of force, and matter had been broken down into its ultimate constituents. After this time, there were two forces instead of one. As I'm fond of telling my students, it's all been downhill since Planck time.

TIMES OF CHANGE FOR THE UNIVERSE

Time from beginning	What happened
10^{-43} seconds	gravity separated from other forces
10^{-33} seconds	strong force separates
10^{-10} seconds	weak and electromagnetic forces separate
10 microseconds	quarks combine to form particles
3 minutes	nuclei of light atoms form
500,000 years	atoms form

Evidence for the Big Bang

932 The first compelling evidence for the Big Bang, (other than the universal expansion itself) was the discovery of the cosmic microwave background. This was discovered in 1964 by two scientists at Bell Laboratories—Arno Penzias and Robert Wilson. Using an early microwave receiver, they set out to measure the microwave radiation coming to the earth from space in order to provide a data base which could be used in the then nascent communications industry. They found that no matter which way they looked, there was microwave radiation coming to the earth (they heard it as a static-like hiss in their apparatus).

Today we interpret this so-called cosmic microwave background as an "echo" of the Big Bang. It is the radiation that was released when atoms formed (see the foregoing) but which has cooled off since that time. It is now characteristic of an object whose temperature is roughly 3 degrees above absolute zero. Such radiation is characteristic of an expanding, cooling universe and is therefore considered to be strong evidence for the Big Bang.

The cosmic microwave background is amazingly uniform. No matter which direction you look into space, the radiation is the same to an accuracy of 0.1 percent. *Why* the radiation should be so smooth remains one of the great problems of cosmology.

933 Nucleosynthesis is another piece of evidence for the big bang. This evidence comes from calculations of the formation of light nuclei at the three-minute mark (see the foregoing). The idea is this: we know from laboratory experiments the probability that a particular product will be produced when two light nuclei collide. For example, if two protons come together at a particular velocity we can measure the probability that they will produce a nucleus of deuterium (a stable isotope of hydrogen). Putting these measured quantities together with the velocities that we expect nuclei to have when the temperature of the universe is as it was at the three-minute mark, we can calculate the number of nuclei of various atoms that should have been produced then.

We can then measure the amounts of each of the Big Bang nuclei in the present universe (correcting for nuclei created in stars) and compare these measured numbers to the predicted numbers. In most cases the predictions are remarkably unforgiving. For example, the abundance of primordial helium in the universe is about 25 percent, which is in line with the predictions. Were the numbers to be as high as 28 percent or as low as 22 percent, on the other hand, the theory would simply be wrong and there would be no way of salvaging it. Thus, the abundances of nuclei give us a very good test of the Big Bang.

934 The use of nucleosynthesis as a test for the big bang allows you to carry out your studies right here on the earth. Deuterium, for example, is an isotope of hydrogen that has one proton and one neutron in its nucleus.

Deuterium is not released from stars, and therefore all of the deuterium on earth was made in the big bang. You can measure the abundance of deuterium in things like ocean water or a rock, then, and get information about the origin of the universe. Arno Penzias calls this "doing cosmology with a shovel."

Large-Scale Structure of the Universe

935 Matter in the universe is almost all in superclusters. Galaxies are grouped together into clusters, and clusters of galaxies are grouped into superclusters. The Milky Way, for example, is part of what astronomers call the local group. This cluster consists of us, the Andromeda galaxy, one other large galaxy you've never heard of (Triangulum) and about twenty small galaxies that are locked in by the gravitational fields of their larger neighbors. The majority of galaxies are found in such groups, although many are located in clusters that

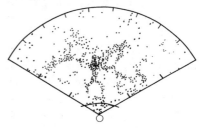

The large-scale structure of the universe. Each dot is a galaxy or cluster of galaxies. The earth is at the apex of the pie-shaped wedge.

may contain a thousand galaxies or more.

Groups and clusters of galaxies, in turn, are found bunched together into superclusters. For example, the local group is one of many such groups in the neighborhood of the Virgo cluster. The Virgo cluster and all the groups like our own make up the local supercluster.

Superclusters are generally long and thin, with clusters of galaxies strung together like beads on a string.

936 Voids are everywhere in the universe. In 1981, astonomers discovered the first of the structures that are now called voids—huge volumes of space with no (or almost no) matter in them. It took a long time to find the voids because it was necessary to sort out the light that shone through the empty region from the darkness associated with the region itself. The first one, located in the constellation Boötes (the plowman) is some 250 million light-years across. Subsequent searches have discovered many such empty spaces in the sky.

Actually, the existence of voids shouldn't seem too surprising. If most of the matter is swept into superclusters, it stands to reason there should be empty spaces somewhere between them.

937 Voids and superclusters together give us our modern picture of the universe. Astronomers are just starting to explore the complex pattern of superclusters and voids that make up what is called the large-scale structure of the universe. The picture that seems to be emerging is simple: imagine taking a knife and slicing through a pile of soapsuds. What you'd see would be a series of empty bubbles, each surrounded by a soap film. Substitute superclusters for the soap film, voids for the bubbles, and voila!—the universe.

938 The universe is made mainly of dark matter. Over the last few years, astronomers have come to realize that 90 percent or more of the matter in the universe is in a form that cannot be seen. A typical galaxy is surrounded by dark matter that makes up more than 90 percent of its mass. In addition, there is evidence that there is extra dark matter in large clusters of galaxies. Dark matter cannot be "seen" in the usual sense, but its presence can be detected by the existence of its gravitational effects.

939 World (?) champion structure The great supercluster of galaxies in the constellations of Perseus and Pegasus is over a billion light-years long, and is the largest supercluster known. In 1989, astronomers at the Harvard–Smithsonian Center for Astrophysics found another structure that they dubbed the "Great Wall." This is a collection of galaxies some 500 million light-years long, 200 million light-years wide, and 15 million light-years thick. There are sure to be even stranger things out there waiting to be discovered—stay tuned!

Enduring Mystery

940 **How did the universe begin?** This problem remains a major issue in theoretical cosmology and in the popular mind. The theories that have been put forward by cosmologists all assume that at some point in the past the constitution of the universe was changed. In the analogy of the bread dough given earlier, you know that if you trace the dough backwards in time you will find it shrinking; but it will never actually shrink to a geometrical point. Instead, the shrinking will go on until you get to the point at which the ingredients of the bread were put together. In the same way cosmologists assume that the time we are calling "zero" does not represent a real physical event, but is merely a backward extrapolation of the present expansion. Theorists differ in terms of what they think was there before the beginning.

The most common theories assume that the universe before the expansion started was unstable and that some event such as the creation of mass set the expansion off.

To appreciate these theories, you have to remember that: (1) mass is a form of energy and can therefore be treated like any other form of energy; and (2) gravitational potential energy can be either positive or negative. These facts mean that it is possible for the positive energy needed to create mass to be balanced by negative energy in the gravitational potential. The creation of matter to form the universe does not necessarily require "something" to be created out of "nothing." It's like digging a hole—when you're done you have a pile of dirt and a hole, but there is no mystery about where the dirt came from. If, however, you could see only the dirt, the digging of the hole would appear miraculous—the dirt would suddenly appear out of nowhere.

In exactly the same way, the creation of the universe looks miraculous because to us it seems that mass suddenly appears out of nowhere. In fact, the matter is just the "dirt" of the universe and is balanced by the "hole" in the form of gravitational fields.

Enduring Mystery

941 **How will the universe end?** The question that seems most obvious to anyone confronted with an expanding universe is whether that expansion will ever end. In the language of the astronomers this is posed as the question, "Is the universe open or closed?" An open universe is one that keeps expanding forever. In a closed universe, the expansion proceeds for a time, then slows down, stops, and reverses itself.

The question of whether the universe is open or closed is, at bottom, an observational one. If there is enough matter in the universe, the gravitational force it exerts will eventually slow down all the outward-rushing galaxies, bring them to a halt, and then pull them back in what astronomers refer to half-jokingly as the "Big Crunch." If there is not enough matter in the universe to do this, then the expansion will go on forever and the universe will be open. A universe which is just on the borderline between being open and being closed is called a "flat" universe.

If you count only the luminous matter—matter that shines and can be seen—the universe contains less than 1 percent of the matter needed to stop the expansion. Thus, it would seem that the universe is open.

Astronomers now believe, however, that the amount of dark matter in the universe is at least 30 percent of the amount needed to stop the Hubble expansion. Many of them believe that more will be discovered in the future and that the universe is, in fact, flat.

It would be nice if this happened, because the best prediction of the total mass of the universe from the inflationary theories is that it is precisely the amount needed to make the universe flat. The prediction is unambiguous—the universe must have this mass to an accuracy of one amoeba per galaxy. For all practical purposes, this means that if the predictions are correct, the expansion of the universe will never reverse and will go on forever.

Enduring Mystery

942 **Why is matter in the universe so lumpy while the microwave background radiation is so smooth?** This is not just *a* question, but *the* question of modern cosmology. The smoothness of the microwave radiation indicates that at one time the universe was uniform, but the universe is far from uniform today. Matter is collected into superclusters and excluded from voids. The problem: how to go from the initial uniformity to the final granularity without producing new microwave radiation that would mess up the observed smoothness?

The Solar System

Introduction

Centuries of observing and decades of working with space probes has produced a lot of information about our own planetary system. After a few remarks about the general structure of the system, we here go on a tour starting from the surface of the sun and extending out through the planets to the farthest comets in the cold of space.

943 **The planets formed at the same time as**

the sun and are made from the same material. The sun and planets formed out of a cloud of interstellar dust about 4.6 billion years ago. Fully 99 percent of all the mass in the interstellar cloud wound up in the sun. The assemblage of planets and other objects in the system outside the sun constitute, on the grand scale, little more than an afterthought.

The rotation of the dust cloud from which the solar system formed eventually spun all the material that didn't go into the sun into a flat disk called the ecliptic. The planets and the rest of the solar system formed in this plane. This explains why all the planets (except Pluto) have orbits that lie in this plane, and all of them move in the same direction (clockwise when viewed from above the north pole). In fact, the solar system looks something like a flat pancake (the ecliptic) with a big cherry (the sun) in the middle.

Gravitational attraction broke the disk of the ecliptic into individual planets. Clumps of material in the disk attracted matter from their neighborhood and, as a consequence, grew more massive. This caused them to attract still more matter and to grow even further. Eventually these aggregated lumps came together to form the planets.

944 **The largest planets in the solar system are least like the earth.** When the solar system was forming and the sun was heating up, there was a crucial difference in temperature be-

tween the inner and outer solar system. Close to the sun where it was warm, a number of materials (such as methane and ammonia) were in vapor form, while farther out they remained in the form of ice. When the nuclear fires in the sun ignited, radiation blew the volatile materials away from the inner solar system, while farther out these materials, along with the gaseous hydrogen and helium, tended to stay around and be incorporated into planets. Thus, planets close to the sun tend to be small and rocky while planets far from the sun tend to be large and gaseous.

The inner, rocky bodies—Mercury, Venus, earth, and Mars—are called the "terrestrial" planets, and the earth's moon is usually included in this category, even though it isn't, strictly speaking, a planet.

The outer planets (Jupiter, Saturn, Uranus, and Neptune) are called "gas giants'" or "Jovian" planets. These planets may have a small rocky core (somewhat larger than that of a terrestrial planet), but they are surrounded by deep layers of liquid and gas.

945 **The Jovian planets probably have a layered structure.** Although we can't get at the interior of these planets, we can make some educated guesses as to what they might be like. One theory holds that each of them has an internal core of metal and rock surrounded by a layer of liquid water, methane, and ammonia. This liquid core is, in turn, surrounded by an outer layer of compressed gases—mainly hydrogen and helium.

946 All the Jovian planets have rings.

The rings of Saturn are one of the best-known features of the solar system, of course, but over the last decade astronomers have discovered that all the other Jovian planets have them as well. Some of these ring systems were first discovered from the earth, but our knowledge of their detailed structure comes from the Voyager satellites (see the following).

947 Pluto is something of an oddball.

For one thing, it's very small—only about 2 percent the mass of the earth—even though it is in the region of the solar system where you might expect gas giants to form. Some astronomers have suggested that Pluto is actually a captured comet, although the fact that the planet has its own moon seems to me to militate against this point of view. How could you capture both a planet and a moon at the same time? Whatever Pluto is, though, it's not like anything else in the solar system.

948 Pluto is not the farthest planet from the sun for part of its "year."

The orbit of Pluto carries the planet inside the orbit of Neptune for part of its "year." Until 1999, in fact, Pluto will be closer to the sun than Neptune.

949 The way a terrestrial planet looks now depends mainly on its size.

As the planets condensed out of the original dust cloud, they took in large amounts of radioactive materials. These materials would have been spread pretty evenly throughout the body of the planet. As they decayed, they generated heat that, when added to the heat created when the planet formed, had to be carried to the surface of the planet and dumped into space. If the planet is small, it has relatively small amounts of radioactive materials, hence small amounts of heat are generated. This heat can be carried to the surface by conduction and radiated away. Mercury, the earth's moon, and Mars are planets like this. When small planets solidify, they acquire a rigid, unchanging outer crust. For a larger planet, there is too much heat to get to the surface by conduction, so convection sets in. (See the section on Plate Tectonics.) Only the earth among the terrestrial planets is big enough to generate this much heat and produce tectonic activity.

950 If the earth were a little smaller, we might not be here.

If the kind of variability in climate that is induced by plate tectonics is important in the evolutionary process, then it is quite likely that intelligent life exists on the earth only because of an accident—the accident that makes the earth a little larger than Venus. If this is true, then the chances of finding extraterrestrial beings elsewhere in the universe is significantly smaller than most people think.

The Outer Regions of the Sun

951 **Looking down into the sun is a little like looking into a dirty pond of water.** You can see about 200 miles into the interior, but that's all. This outer part of the main body of the sun—what would correspond to its surface if it were solid—is called the photosphere (sphere of light). Above the photosphere are two more layers—the chromosphere (sphere of color), which extends a few thousand miles above the top of the photosphere, and the corona (or crown). Most of the activity that we associate with the sun—sunspots, prominences, and solar flares (see the following)—occurs in the photosphere and chromosphere. The chromosphere gets its name from the fact that it contains hydrogen gas, which emits a red light that was picked up by solar telescopes in the nineteenth century.

Enduring Mystery

952 **Why is the corona so hot?** Contrary to what you might expect, the corona is very hot—much hotter than the photosphere. Not only that, it gets hotter the farther away from the surface you go, reaching temperatures of several million degrees at its maximum. This is puzzling—it's as if the air at the top of Mount Everest were hotter than air at sea level.

One guess at a solution to this puzzle is the theory that convection cells at the surface of the sun churn around and create sound waves, which then move out into the chromosphere and heat it. You can think of them as being somewhat analogous to sonic booms. Another theory is that the heating results from the dumping of energy from magnetic fields.

953 **Even though the corona is at an extremely high temperature, you could stick your hand into it and never feel a thing.** The atoms in the corona have a lot of energy, so much so that most have been stripped of most of their electrons. The atoms are so sparsely scattered, though, that relatively few would hit your hand. The total energy in the form of heat you'd absorb would be small.

954 **The sun's face is marked by prominences and solar flares.** If you've seen closeup pictures of the sun, the most striking thing you saw was probably the great arches and loops of flaming material that rose and fell back down on the surface. These so-called prominences typically occur in regions where there are sunspots and are related to activity in the sun's magnetic fields. They generally last a few days.

Occasionally, a prominence explodes and spews large amounts of material into space. This is called a solar flare. When particles from a major solar flare hit the earth's atmosphere, they can disrupt radio communication for a period of several days.

---

955 **Solar flares are a major environmental hazard faced by astronauts.** One of the greatest worries during the Apollo missions, for example, was that once the astronauts were irrevocably committed to lunar orbit a major flare would erupt on the sun, sending lethal doses of radiation out toward the unshielded spaceship.

---

956 **The sun's magnetic field is nothing special.** In fact, on the average it's only a few times stronger than that at the surface of the earth, though it varies erratically over the solar surface.

The interesting thing about the sun's magnetic field is that it extends far out into space. In fact, you can think of the sun's magnetic field as being like a fabric that ties the entire solar system together. Within this all encompassing magnetic field (and continuously linked to it) are embedded the magnetic fields of each of the planets.

957 **The sun is constantly emitting particles that flow outward through the solar system.** We call this flow the "solar wind." You can think of the particles of the solar wind as sliding out along the lines of the solar magnetic field like beads sliding along a wire.

As it blows by each planet, the solar wind distorts the planetary magnetic field so that it is compressed on the "upstream" side and stretched out on the "downstream" side. Below is a sketch of the earth's magnetic field (or magnetosphere) as it is distorted by the solar wind.

Astronomers often describe magnetic fields in the terms you might use if there were water flowing through the solar system. For example, the bunching up of the magnetic field around the planet is called the "bow wave," an analogy to the water that bunches up in front of a moving boat.

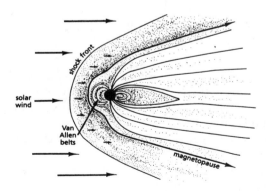

The earth's magnetic field as distorted by the solar wind.

The Individual Planets

In the following entries, we give a thumbnail sketch of the important facts about each of the nine planets in the solar system. A table with relevant numbers is appended at the end of the discussion.

958 **Mercury.** The closest planet to the sun, Mercury completes one trip around its orbit every eighty-eight days. The planet is visible from the earth as a morning and evening star. It used to be thought that, in analogy with the earth's moon, Mercury always kept the same face toward the sun. There were many wonderful science-fiction stories written following that theme. Unfortunately for the authors, the planet does rotate—each face gets its share of sunlight.

Mercury has no atmosphere. Its surface is pitted with craters and looks very much like the surface of the earth's moon. The planet has an interior somewhat like that of the earth, with a metal core surrounded by a layer of silicon-based minerals. Unlike the earth, however, Mercury's entire structure is frozen solid, and does not move.

959 **Venus.** Venus is the planet most like the earth. The temperature at its surface is high—some 470 degrees centigrade (800 Fahrenheit). It is believed that the reason for the high temperature is a greenhouse effect caused by the large amounts of water vapor and carbon dioxide in the Venusian atmosphere. The planet is swathed in multiple layers of clouds, so detailed information about its surface has traditionally been hard to come by.

There used to be wonderful science-fiction stories about the swamps of Venus—it was believed at one time that Venus was like the Florida Everglades, only more so. Given its surface temperature, this is now considered to be impossible. We know a great deal about the surface conditions on Venus because Soviet spacecraft have landed there and sent back both pictures and data.

Today, orbiting satellites have given us radar maps of the Venusian surface. Although there are volcanoes on the surface, there appears to be little tectonic activity.

Pop Quiz

Why do Venus and Mercury only appear as morning stars? Answer: Because they can never be in back of the earth, which is where they would have to be to be seen in the middle of the night.

960 **The earth-moon system.** The earth is the only planet in the solar system that has tectonic activity, the only planet that has liquid water on its surface, and the only planet that contains life.

The earth's moon is the only body in the solar system whose features we can detect with the naked eye. The "man in the moon" is familiar to us all from childhood. The dark areas of the moon, called *mare* (from the Latin word for "sea"), are actually old flows of basalt. It is believed that they were formed when the moon had cooled enough so that it had a crust, but was still molten underneath. The impact of a large meteorite could pierce the crust and initiate a large-scale lava flow. It is this solidified lava that we see in the luna mares today.

In addition, the moon has highlands which are made up of rings of craters. The moon has no mountain chains of the type known on the earth.

Instruments left behind by the Apollo astronauts during their missions to the moon allowed us to measure "earthquakes" (moonquakes?) and therefore to form a pretty good picture of what the interior of the moon is like.

Even though we don't know exactly how the moon formed, we do know that it must have formed about the same time as the earth. Therefore, the dating of moon rocks brought back by the Apollo astronauts is an important way of determining, not only the age of the moon, but of our planet and of the entire solar system. That age, of course, is now believed to be 4.6 billion years.

961 **Mars.** The farthest out of the terrestrial planets, Mars is only about half the size of the earth. Its "year" is two earth-years long, and we can tell that it has seasons because we can watch the polar caps wax and wane. The polar caps of Mars are not made of water but of frozen carbon dioxide (dry ice). We can also see weather on the surface of Mars in the form of massive dust storms. It is believed that, although there's no water on the surface now, there may have been water on the surface at one time. (The evidence for this is the existence of erosional features that look a lot like river valleys formed by running water.)

The surface of Mars as seen by the Viking Lander.

962 **The Viking missions gave us our best information on Martian conditions.** On July 20, 1976, the American probe Viking I landed on Mars, the first manmade object to visit another planet. It was followed quickly by Viking II. Beside giving us unforgettable photographs of the pink skies of Mars, the landers' chemical experiments searched for evidence of life on our sister planet. At present, there are no results from these experiments that cannot be explained in terms of simple chemical reactions between nonliving substances.

963 **The largest mountain in the solar system is on Mars.** Olympus Mons, an extinct (?) volcano, is 17 miles high and over 370 miles across at the base—big enough to stretch from Boston to Baltimore. A volcano this big on earth would collapse under its own weight, but in the smaller Martian gravity it stands tall.

964 **The canals of Mars aren't really there.** In the 1880s, the Italian astronomer Giovanni Schiaparelli reported what he called "canali" on the surface of Mars. The word means "channels" in Italian but it was translated into English as "canal," a word which unfortunately carries a connotation of intelligent design. In the early part of the century, the American astronomer Percival Lowell (1855–1916) did extensive studies in which he reported that the Martian surface was crisscrossed with canals. He even watched the color of these canals change from brown to green as the Martian seasons progressed and calculated the rate at which the vegetation moved south along the canals—to two decimal places, no less! All this despite the fact that

965 **There aren't any faces on the surface of Mars either.** Recently, supermarket tabloids have headlined a "face" of Mars—a couple of low hills which, in a certain light, look like a human face. You can judge the seriousness of these claims by noting that the most recent headline I saw claimed the face was that of Elvis Presley.

966 **What did Lowell see?** I recently had a letter from Clyde Tombaugh, one of the grand old men of the astronomical community and the discoverer of the planet Pluto (see the following). Tombaugh explained that the reason Lowell saw canals had to do with the details of the way he set his telescope—it made unconnected dots look like a line. Thus, he wasn't necessarily imagining things, and may only have been the victim of injudicious calibration of his instruments.

967 **There is no evidence for life on Mars or any other body in the solar system.** On Venus, the moon, and Mars, the three bodies we've visited (in person or by probe), there is no evidence for the existence of life. This would have come as quite a surprise to scientists as late as the 1950s, when it was felt that the two planets, at least, probably harbored life. Today, those who want to believe that life is common in the universe urge expeditions to Mars to look for fossils. The idea is that life (like liquid water) once existed there, but does so no longer.

968 **The asteroid belt.** The asteroid belt is a large collection of loose material, ranging in size from pebbles to bodies several hundred miles across, that exists in the space between the orbits of Mars and Jupiter. Contrary to what *Superman* comics would have us believe, the asteroid belt was never a planet that broke into pieces. Instead, it is a planet that never quite made it. Scientists believe that the gravitational influence of Jupiter prevented the asteroids from forming into a planet.

969 **Jupiter.** The largest planet, Jupiter, rotates rapidly—its "day" is only ten hours long. As far as we can tell, the density of Jupiter gets higher and higher as you move down into the body of the planet, but standing on it would be roughly like trying to stand on a milkshake—none of the Jovian planets have a "surface" in the sense of "solid ground."

Because of its rapid rotation, the atmosphere of Jupiter is split into different-colored bands. Deep inside the planet, as just discussed, there may be a small, earth-sized rocky core.

Jupiter, with the "great red spot" on the lower right.

970 **The moons of Jupiter.** Jupiter has many moons, all of which circle the giant planet much as the planets circle the sun. The five brightest of these moons are called the "Galilean satellites" because they were first seen by Galileo when he turned his telescope to the sky. The number of known moons of Jupiter has increased steadily as our powers of observation have gotten better—the current count being sixteen.

Many of the moons of Jupiter are quite large and resemble the terrestrial planets in their makeup. Some salient features: Callisto and Ganymede are probably made up primarily of water ice, and the surface of the former is the most cratered object in the solar system. Io has been observed to have active volcanoes and, hence, is a favorite object of study by geologists.

When the Hubble space telescope is working, we will be able to have photographs of the moons of Jupiter equivalent to those that were taken by the Voyager satellites in their flyby in 1979.

971 Looking at Jupiter through a telescope, you can see what looks like a large red eye that moves around with the planet as it rotates. This so-called "great red spot" is a gigantic hurricane that has been going on for as long as we have been observing the planet. It is clearly the most destructive weather (if that's the right word) that we know of—you could drop the entire United States into it without causing a ripple.

972 The volcanoes of Io, first seen by the Voyager spacecraft, spew out sulphur that constantly coats the surface of the moon. Every square foot of Io's surface is less than a thousand years old!

973 Jupiter was almost a star.

The mass of Jupiter is only eighty times less than would be needed to raise its internal temperatures to the point where the fusion reaction could ignite. Thus, the solar system narrowly missed having two stars. Had Jupiter ignited it is very unlikely that life would have developed on earth, since the extra radiation from even a small star that close would have upset the delicate balance that makes life possible on our planet.

974 Saturn.

Saturn with its rings is easily the most spectacular of the planets. A gas giant like Jupiter, Saturn is also the last of the planets that can be seen from earth with the naked eye. It has twenty-one moons and one of them—Titan—is the largest satellite in the solar system—the only satellite that possesses an atmosphere (it consists of nitrogen, methane, and argon). The surface temperature of Titan hovers around −280° C (958 K). This combination makes Titan somewhat similar to the earth before life developed, and the surface may well have all sorts of organic molecules floating in liquid methane.

The rings of Saturn probably attract more attention than anything else about the planet. They are narrow bands of debris, mostly in the form of rocks and ice. There seems to be some disagreement among astronomers at the present time as to whether they represent the remains of a satellite that was torn apart by Saturn's gravitational force or whether, like the asteroid belt, they

Saturn with its rings.

are material from a satellite that never quite made it.

---

975 **The rings of Saturn are very thin.** Some astronomers think that the rings of Saturn, although they reflect a lot of light, may be no more than a few hundred feet thick. The entire celestial display, in other words, may take up no more vertical space than a moderate-sized downtown building.

976 **Uranus.** On March 13, 1781, the astronomer William Herschel saw what he called "either a nebulous star or a comet" in his telescope while he was living in the town of Bath, England. This later turned out to be the planet Uranus, and Herschel achieved lasting fame as the first human being to discover a new planet. Like the other Jovian planets, Uranus probably has a three-layered structure. The planet may not be big enough to compress the internal layer to a solid, however; some astronomers think that the core of Uranus may be nothing more than a thick, viscous fluid.

Uranus has five moons and has a series of very narrow, dark rings around it, somewhat like the rings of Saturn. These were discovered in 1977 when the planet passed in front of a star and the diminution of light due to absorption in the rings was seen.

977 **Uranus rotates on its "side."** Most of the planets in the solar system rotate around an axis so that over the course of a day both sides are exposed to the sun. Unlike them, Uranus lies on its side so that its axis of rotation lies in the same plane as its orbit. Thus, the south pole gets light for half of the "year" and the north pole gets light for the other half.

Pop Quiz

Where would you see the sun rise if you lived on the north pole of Uranus? In the south.

---

978 **The interplanetary magnetic field had nothing to do with manna.** One of the joys of teaching is that students often introduce you to ideas you would never have encountered on your own. After being asked several times about the writings of Immanuel Velikovsky, I read his *Worlds in Collision* in self-defense. This is a strange theory about the recent history of the solar system, which includes Venus being torn from Jupiter and raining manna on the children of Israel as it passed the earth on its way to its present orbit. This is all supposed to be related in some way to magnetic forces, and the picture of the magnetic web I've just given you is frequently cited as evidence that it all *could* have happened that way. Unfortunately, the forces associated with the interplanetary magnetic field are billions of times too small to move planets around.

Besides, how did Venus know not to rain manna on the Sabbath, as the Bible story states?

979 Neptune. The August 25, 1989, flyby of Neptune removed a good deal of the mystery which used to shroud this planet, then created new puzzles. Like the other Jovian planets, Neptune has eight known moons and its own set of rings. The winds on its surface are the fastest in the solar system, clocking in at up to fifteen hundred miles per hour. It has its own long lasting storm, similar to Jupiter's great red spot, called the great dark spot.

980 Neptune was the first planet that was discovered as the result of a prediction. Noting deviations in the orbit of Uranus from its predicted course, astronomers in the nineteenth century calculated how big and where a planet would have to be to cause those deviations. They directed their telescopes to that spot and made the discovery of the planet on September 23, 1845.

981 The Voyager missions taught us more about the outer planets than any scientific project ever undertaken. Launched in August and September 1977, Voyager I and II were intended to take advantage of a rare alignment—an alignment that would allow a single spaceship to visit all but one of the Jovian planets in one shot. Voyager I ran interference, visiting Jupiter and Saturn a few months before its more heavily instrumented sister and giving scientists a warning about what they would encounter. The two flew by Jupiter in 1979, Saturn in 1980–81, and Voyager II continued on to Neptune for an encounter in August 1989. The box score for the mission: the Voyagers visited three gas planets, twelve large moons, three ring systems (containing thousands of individual rings), and sent back 5 trillion bits of data (including 100,000 pictures). Voyager II will leave the region of our sun's magnetic influence in about twenty years. In the year 8571 it will pass Barnard's Star and in the year 296036 it will pass Sirius, the brightest star in our sky.

982 Pluto. Pluto is in many ways the strangest of the planets. It is small and has a large moon (called Charon). Its orbit is eccentric, which may cause it to have seasons in the sense that when it is close to the sun, the liquid methane on its surface boils to form a kind of atmospheric haze. When the planet moves farther away from the sun, it starts to snow solid methane.

983 Pluto is not dark. Despite its great distance from the sun, the surface of Pluto is probably as bright as a moonlit night on earth. The reason is all that methane, which is as white as newly fallen snow.

984 The discovery of Pluto was more accident than design. The American astronomer Percival Lowell had predicted the existence of a ninth planet (he called it Planet X) based on what he took to be irregularities in the orbit of Neptune. Today astronomers argue that these "irregularities" weren't real, but the result of instrumental error. Nevertheless, Lowell produced predictions about where Planet X ought to be (although, to be honest, the predictions changed occasionally when he redid the calculations). In any case, in 1930 Clyde Tombaugh, doing a systemic sky survey that would have found the planet no matter where it was, discovered the planet we now call Pluto. By coincidence, its position was pretty close to where Lowell's last prediction said it should be. Was it just luck? We'll never know.

985 When Pluto was first discovered, it was assumed that it had a mass about the same as that of the earth. In the time since then, scientists' estimate of its mass has steadily decreased. At a talk I attended recently one astronomer showed a graph in which the estimated mass of Pluto was plotted as a function of time. His point was that if you just drew a line through all the data points, it would appear that the mass as estimated by humans should become zero sometime in the 1980s. This has resulted, he told us, in a paper titled. "On the Impending Disappearance of the Planet Pluto."

DISTANCES

If the earth were a ten-pound basketball located one hundred feet from the sun:

Mercury is a half-pound baseball forty feet from the sun
Venus is an eight-pound basketball seventy feet from the sun
Mars is a one-ounce cantelope one hundred and fifty feet from the sun
Jupiter is a twenty-foot truck five hundred feet from the sun
Saturn is a small subcompact car one thousand feet from the sun
Uranus is a hundred-and-fifty pound sofa two thousand feet from the sun
Neptune is a little heavier than Uranus, but over half a mile from the sun
Pluto is a quarter-ounce baseball four thousand feet from the sun

and their "days" and "years" are approximately

Planet	"Day"	"Year"
Mercury	59 earth days	3 earth months
Venus	243 earth days	7 earth months
Mars	1 earth day	1 earth year, 10.5 earth months
Jupiter	10 hours	12 earth years
Saturn	10 hours	29.5 earth years
Uranus	1 earth day	84 earth years
Neptune	1 earth day	165 earth years
Pluto	6 earth days	248 earth years

Notice that since its discovery, Pluto has had time to cover only about 20 percent of its orbit, and that the last time it was in its present position was before the Revolutionary War.

986 **The Oort cloud.** Far outside the orbit of Pluto, a large collection of dirty snowballs (potential comets) circles the solar system. They form a cloud named after the Dutch astronomer Jan Oort, who first suggested its existence. Occasionally one of these snowballs is jostled loose and enters the inner solar system as a comet. Oort deduced the existence of the cloud by tracking the orbits of incoming comets backward to their point of origin. The Oort cloud figures prominently in some theories of mass extinctions on earth.

Meteors and Meteorites

987 **Stuff is falling into the earth from space all the time.** Rocks of all sizes—from grains of sand to mountains—have fallen and continue to fall to the earth. Because of this infall, the earth is getting heavier at the rate of twenty tons per day.

The common "shooting star," or meteor, is a track of light in the sky that occurs when a piece of space dust about the size of a grain of sand burns up in the atmosphere. If the object is this small, it will burn up completely before it gets very far into the atmosphere. Larger objects make larger tracks, and may even survive the fall and hit the ground.

988 **If a falling body makes it to the ground, it's called a meteorite.** A larger object can have its outer skin burned off and still have enough material left to make it to the ground. Such meteorites can be seen in museums, where they are easily recognized by the black, burnt appearance of their outer skin.

Meteorites have been found from the size of gravel to objects that weigh tens of tons.

989 **Scientists were historically reluctant to accept the notion that objects could fall from the sky.** For example, Edmund Halley argued in 1718 that a spectacular shooting star that had been seen over most of Europe was in fact a conflagration of "inflammable sulfureous Vapors" in the upper atmosphere. Even Thomas Jefferson, upon being informed that two professors from Yale had confirmed the fall of a meteorite in New England, is supposed to have remarked, "I would rather believe that two Yankee professors would lie than that stones fall from the sky."

990 **Meteor Crater in Arizona was caused by the impact of a large body.** The crater near Flagstaff, Arizona is probably the best-known evidence for the impact of a large body on the earth. About 25,000 years ago, an iron rich rock weighing about

Meteor crater in Arizona.

10,000 tons hit the plains of Arizona, digging out a crater more than a mile across. As impacts go, this was a relatively mild event.

991 The Tunguska event in Siberia was caused by a meteorite. In 1908, a meteorite entered the atmosphere over Siberia, broke up into pieces, and hit the ground near the Tunguska river. The crash flattened the forest for hundreds of miles and sent shock waves through the air all around the world. Subsequent studies have turned up bits of melted rock from the meteorite.

The Tunguska event has a separate existence in folklore, where it has been blamed on everything from a volcanic eruption to the impact of a black hole. Its origin, alas, is much more prosaic.

992 There are over 120 known meteorite craters on the earth. Geologists have discovered over 120 places on the earth where meteorites have hit in the past. These range from structures a hundred miles across to holes a few hundred feet across with the original meteorite still at the bottom.

In fact, it's a good bet that most craters haven't been discovered yet. For one thing, a meteorite hitting the earth has roughly three chances in four of hitting the ocean. The ocean floor isn't well explored, and there's no telling what's waiting for us there. Even on land, old meteorite craters—particularly large ones—are rather hard to recognize. They may have been big craters when they were formed, but like any big hole in the ground, they would have filled up with water and become lakes that gradually silted in. Today, these craters are simply a circular range of hills twenty or thirty miles across, with flat level land in between. For example, Manacougan crater in Quebec, is marked only by a ring of lakes sixty miles across, and is readily recognizable only from satellite photos.

We know intellectually that 'the earth is part of the solar system— that our planet is an integral piece of the entire assembly. Having a concept in your mind, however, isn't the same as being reminded of a fact by having a mountain-sized rock dropped on your head every few hundred million years. The existence of large craters reminds us that the earth is not isolated in space.

993 Just as scientists were reluctant to accept the idea that objects could fall from the sky, they were reluctant to accept the idea that objects falling from the sky could leave a scar on earth. Until the 1950s, for example, Meteorite

Crater in Arizona was blamed on a "gas bubble" coming up from inside the earth. One meteorite fall in Argentina that left a hole with the meteorite at the bottom was blamed by a geologist on "prehistoric Indians who dug the hole and then buried the sacred iron object there."

994 During the early history of the solar system, the rain of meteorites was much heavier than it is now. Just after the planets were formed, there was a lot of space junk orbiting the sun. Consequently, during this time the rain of objects onto the planets (including the earth) was much heavier than it is today. Astronomers sometimes speak of the Great Bombardment when they refer to this period. We estimate that for the first billion years in the life of the planets, meteorites, both large and small, fell at a very high rate. We can still see the craters from this bombardment on the moon, Mercury, and other airless objects in the solar system.

995 Although the craters on the moon are still there, those on the earth have weathered away. Presumably very large craters were made early on in the history of the earth. Since that time, however, the forces of erosion and weathering have destroyed almost all of the early craters that existed on the surface of the earth. Consequently, our planet (along with Mars and Venus, which also have weather) seems to be largely devoid of the craters, even though airless (and hence weatherless) bodies like the moon and Mercury have them in abundance.

996 Meteorites carry important clues to the origin of the solar system. In fact, because they were not taken up into planets when the solar system formed, meteorites are a museum of the materials from which the sun and the planets were made. They have been floating unchanged in space for billions of years, and when they fall to earth they bring with them information about the beginning. Because of this, they are avidly studied by chemists and geologists. The idea is that if we know what the earth is like now and how it started out, we should be able to figure out what happened in between.

997 Some meteorites come from the asteroid belt, others are burned out comets. Occasionally, bits of rock in the asteroid belt collide with each other, knocking one or more of the asteroids onto a path that brings it inside the orbit of the earth. A piece of rock in such an orbit is called an Apollo asteroid. Every once in a while, the laws of chance dictate that some of these objects will strike the earth.

A second major source of meteorites are the comets in the Oort cloud. When comets first come into the solar system, the heat of the sun boils off all the materials that can be boiled, so that after many passes a comet becomes burnt out and only the rocky core remains. Astronomers estimate that about half of the large objects whose orbits could bring them into collision with the earth are burnt-out comets.

998 One of the most astonishing discoveries in the last few years is that there are occasional meteorites which began life either on Mars or on the moon. The idea was that there was a large impact on the other body, which kicked material back into space. This material then went into orbit like any other piece of space junk, and eventually it fell to earth. At the moment, there are fewer than half a dozen verified pieces of rocks of this type.

In other meteorites, scientists have found small grains of material (mostly diamonds) that seem to be different from the stuff that makes up the solar system. The explanation is that these grains were created in a supernova long before the birth of the sun, drifted through space, and were then incorporated into the stuff from which the planets were made.

999 One of the best places to look for meteorites is in Antarctica. Meteorites usually tend to explode on impact or stay where they fall, burying themselves deep into the earth. Occasionally, however, one falls on one of the great ice sheets in Antarctica. Once it's buried in the ice, it is carried along by the flow of the glacier. There are a couple of places in Antarctica where the ice is pushed up the side of the hill. When the ice gets to the top, the blowing wind "evaporates" the ice. When the ice is gone, the meteorites are left behind, and scientists can walk along the ridges and pick them up. They are, in effect, using the entire antarctic ice sheet as a collector and conveyer belt to bring them the meteorites. Many of the unusual meteorites like the ones that come from Mars or the moon were discovered in these kinds of regions.

1000 Meteorites are the ultimate mineral resource for the human race. When the earth was formed, it went through a period when it was molten, and many heavy materials (like iron) sank toward the center. What's left at the surface—what we mine and use—is more or less the tail end of this great reserve. Asteroids, since they were never incorporated into a planet, never went through this process. Consequently, asteroids are very rich in materials that are valuable at the surface of the earth. They abound in iron, nickel, cobalt, gold, and other heavy metals. A number of visionaries have pointed out that there is enough material in a single asteroid to last the human race for hundreds of years, even at present rates of consumption. My own calculations indicate that the market value of the minerals in an asteroid several miles across is probably several trillion dollars. Sometime in the next few centuries, the human race will discover this immense resource floating around above our heads and begin to use it. One good effect of this discovery will be the end of strip-mining and the destruction of the habitat here on earth.

1001 You can learn more about science on your own. This is probably the most important thing you should know about science. To get you started, here are a few books I have enjoyed (with apologies to the authors of the many wonderful books I can't include because of space limitations).

Biology

Charles Darwin. *On the Origin of Species.* New York: Penguin Classics, 1986 (reprint).
Larry Gonick and Mark Wheelis. *Cartoon Guide to Genetics.* New York: Bantam, 1985.
Stephen Jay Gould. *The Panda's Thumb.* New York: W. W. Norton, 1982.
James D. Watson. *The Double Helix.* New York: Atheneum, 1985 (reprint).

Physics and Chemistry

P. W. Atkins. *Molecules.* New York: Scientific American Library, 1987.
J. E. Gordon. *The New Science of Strong Materials or Why things Don't Fall Down.* Princeton, N.J.: Princeton University Press, 1976.
Heinz R. Pagels. *The Cosmic Code.* New York: Simon & Schuster, 1982.
Michael Riordan. *The Hunting of the Quark.* New York: Simon & Schuster, 1987.

Earth Sciences and Astronomy

Williard Bascom. *Waves and Beaches.* Garden City, N.Y.: Doubleday/Anchor, 1964.
Timothy Ferris. *Coming of Age in the Milky Way.* New York: William Morrow, 1988.
John McPhee. *Basin and Range.* New York: Farrar, Strauss, and Giroux, 1981.
————. *Rising from the Plains.* New York: Farrar, Strauss, and Giroux, 1986.
————. *Control of Nature.* New York: Farrar, Strauss, and Giroux, 1989.

Photo Credits

1, Gregory G. Dimijian, M.D./Photo Researchers; **5,** R. J. Erwin/Photo Researchers; **7T.** Tom McHugh/Photo Researchers; **7B,** Fred Bavendam/ Peter Arnold, Inc.; **8B,** Al Giddings/Ocean Images; **8T,** Tom McHugh/ Photo Researchers; **12,** David Schart/Peter Arnold, Inc.; **15T,** Bloom & Fawcett/Photo Researchers; **15B,** Bio Photo Assoc./Photo Researchers; **18B,** David M. Phillips/The Population Council/Science Source/Photo Researchers; **20,** Andrew Martinez/Photo Researchers; **22,** Francis Leroy/ Biocosmos/Science Photo Library/Photo Researchers; **23T, B,** Hans Pfletschinger/Peter Arnold, Inc.; **29,** M. I. Walker/Photo Researchers; **40,** Lawrence Migdale/Photo Researchers; **47,** Stephen Dalton/Photo Researchers; **50,** Mary Evans Picture Library/Photo Researchers; **51,** George Holton/ Photo Researchers; **53,** Kobal Collection/Superstock; **60,** John Reader/ Science Photo Library/Photo Researchers; **63,** Field Museum of Natural History, Chicago, #59442; **73,** CDC/Science Source/Photo Researchers; **90,** Dr. R. L. Brinster/Peter Arnold, Inc.; **100,** A. J. Belling/Photo Researchers; **106,** Ed Reschke/Peter Arnold, Inc.; **109,** The Bettmann Archive; **115,** Jan Robert Factor/Photo Researchers; **126,** The Bettmann Archive; **142, T, B,** George Whiteley/Photo Researcher; **152,** Santi Visalli/The Image Bank; **158,** UPI/Bettmann; **161,** Fermilab; **170,** The Bettmann Archive; **176,** Phillip Hayson/Photo Researchers; **177,** Bell Labs; **179,** Omikron/Science Source/Photo Researchers; **211,** NASA; **224,** Paolo Koch/Photo Researchers; **226,** Francois Gohier/Photo Researchers; **239,** Max & Kit Hunn/Photo Researchers; **247,** Goddard Space Flight Center/NASA; **255,** David Parker/ Science Photo Library/Photo Researchers; **275,** Finley-Holiday Films; **277,** Finley-Holiday Films; **278,** Finley-Holiday Films; **283,** Galen Rowell/Peter Arnold, Inc.

Photo Research by Julie Tesser

Diagrams by Judith Peatross

Index

About the Author

JAMES TREFIL, Clarence J. Robinson Professor of Physics at George Mason University, is author of more than one hundred professional papers, three textbooks, and twelve other books on science. He has been a commentator on National Public Radio and is a Fellow of the American Physical Society. He also serves on the Committee on Fundamental Constants and Basic Standards of the National Research Council. A former John Simon Guggenheim Fellow, Trefil has won the AAAS Westinghouse Award for excellence in science writing, and his teaching has been recognized by the Innovation Award of the National University Continuing Education Association/American College Testing Programs. His books include *The Moment of Creation, A Scientist at the Seashore, Meditations at 10,000 Feet, The Dark Side of the Universe, Reading the Mind of God,* and, as coauthor, *The Dictionary of Cultural Literacy* and *Science Matters: Achieving Scientific Literacy.*